T0235894

INTERNATIONAL CENTRE FOR MECHANICAL SCIENCES

COURSES AND LÉCTURES - No. 55

GERHARD SCHWEITZER
TECHNICAL UNIVERSITY OF MUNICH

CRITICAL SPEEDS OF GYROSCOPES

COURSE HELD AT THE DEPARTMENT
FOR GENERAL MECHANICS
SEPTEMBER 1970

UDINE 1970

Springer-Verlag Wien GmbH

© 1972 by Springer-Verlag Wien
Originally published by CISM, Udine in 1972.
ISBN 978-3-211-81150-4   ISBN 978-3-7091-4334-6 (eBook)
DOI 10.1007/978-3-7091-4334-6

Preface

The lecture notes presented here are based upon a course held at the International Centre for Mechanical Sciences, Udine, in Fall 1970. The objectives of the course are to evaluate the dynamic behavior of high-speed rotors and to show methods for calculating, optimizing and measuring this behavior. As the course partly deals with the theory of gyroscopic systems it relates to the course of Prof. K. Magnus on "Gyrodynamics" held during the same session.

I would like to express my gratitude to CISM, in particular to Prof. L. Sobrero, the Secretary General of the Centre, for his kind invitation and I wish to express my sincere thanks to Prof. K. Magnus for his continuous interest and support.

Udine, September 1970   G. Schweitzer

Contents

1. Introduction

The term "gyroscope" is defined here as a high-speed rotor with moments of inertia that can not be neglected so that the rotor is subject to gyroscopic forces. High-speed rotors are of increasing technical interest. They are used in turbines, grinding-machines, centrifuges and gyroscopic devices of various constructions. The design of these rotors requires knowledge and control of their dynamic behavior and motion. The critical speeds characterize an essential part of the dynamic behavior. They denote the speeds for which the flexural vibrations reach dangerous values and lead to a critical state. Many of the questions arising thereby have already been treated in numerous publications. The investigations of SMITH [1] and the works of BIEZENO-GRAMMEL [2] shall be mentioned in particular. The increasing technical requirements and the now available theoretical and experimental means demand and permit more extensive investigations, particularly with regard to the optimum of the rotor design that is aimed at in the final analysis.

The lecture is based on publications, listed in the bibliography, and on investigations that have been effected by the Institute of Mechanics, Technical University München, in recent years. The following themes are dealt with : The kinetic behavior of a rotor, mounted on a rotating cantilever, is de-

rived in detail . By this easily calculated example,technical
terms and the concept of further evaluation are introduced
(Chapt. 2). The fundamental equations for small vibrations of
a gyroscopic system are stated. They consist of a system of
linear differential equations, for which general stability criteria
are valid. From the solutions of the homogeneous and   nonhomogeneous system,conclusions about the eigenbehavior and
the behavior against disturbances, about effects as forward
and reverse whirl, resonance and apparent resonance are
drawn (Chapt. 3).

The general results are applied to a more complex sys_
tem; i. e.,to a rotor with elastically supported bearings(Chapt. 4).
For this example methods and results of optimizing the rotor
motion are presented and different performance criteria are
discussed. The experimental aspect of studying the rotor motions is shortly dealt with, considering particularly the experimental over-all concept and data processing.

For a high-speed rotor it is absolutely necessary that
it is balanced in order to avoid large loads on the rotor and
therefore the principle of balancing is explained (Chapt. 5). In
the last chapter different effects are shown that can modify the
general motion of the rotor considerably.

2. Rotor on overhung shaft

2.1. Equations of motion.

      The equations of motion for a rotor mounted on the end of a rotating cantilever are derived in analytically and complex variables are introduced. Physical interpretations of the different terms of the equations are given.

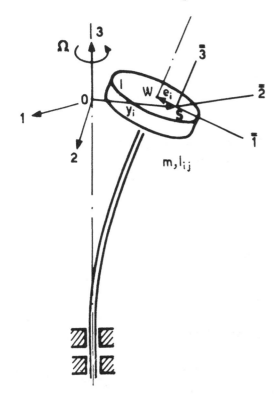

Fig. 1

Fig. 1 shows the notations and coordinate systems. The origin 0 of the fixed 1 2 3 coordinate system coincides initially with the center W of the rotor. The body-fixed $\bar{1}$ $\bar{2}$ $\bar{3}$ coordinate system, with its origin at the center of mass S of the rotor, is directed so that the $\bar{3}$-axis is parallel to the axis of symmetry of the rotor, that the center W lies on the $\bar{1}$-axis and that the vector of eccentricity is $\bar{e}_i = \begin{bmatrix} -e, 0, 0 \end{bmatrix}^T$. Only eccentricity of the center of mass is taken into account. Let the rotor itself be symmetrical, its tensor of inertia is then

$$\bar{\bar{I}}_{ij} = \begin{bmatrix} A & 0 & 0 \\ 0 & A & 0 \\ 0 & 0 & C \end{bmatrix}.$$

The transformation between the two coordinate systems is expressed by the cardanic angles α, β, γ with the transformation matrix c_{ij} of Fig. 2.

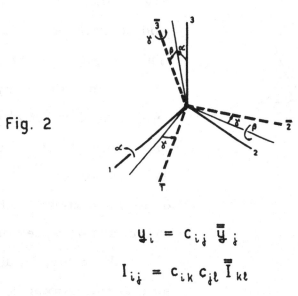

Fig. 2

$$y_i = c_{ij} \bar{\bar{y}}_j$$

$$I_{ij} = c_{ik} c_{jl} \bar{\bar{I}}_{kl}$$

$$c_{ij} = \begin{bmatrix} \cos\beta \cos\gamma & -\cos\beta \cos\gamma & \sin\beta \\ \cos\alpha \sin\gamma + \sin\alpha \sin\beta \cos\gamma & \cos\alpha \cos\gamma - \sin\alpha \sin\beta \sin\gamma & -\sin\alpha \cos\beta \\ \sin\alpha \sin\gamma - \cos\alpha \sin\beta \cos\gamma & \sin\alpha \cos\gamma + \cos\alpha \sin\beta \sin\gamma & \cos\alpha \cos\beta \end{bmatrix}$$

Another set of angles ψ, δ, Φ as used in flight mechanics would be equally suitable here. The angle ψ indicates the rotation on the vertical, δ and Φ are inclinations of the rotor axis with respect to the vertical. Application of these angles can be recommended when the motion of the rotor is studied in a coordinate system rotating with the angular speed of the rotor on the vertical. The Euler angles, however, are not suitable for this problem. The transformation matrix becomes singular and it admits therefore no discrimination between the rotation on the 3- and the $\bar{\bar{3}}$-axis when both axes fall together.

The angular velocity of the rotor with respect to the inertial system is

$$\Omega_i = \begin{bmatrix} 0 \\ 0 \\ \Omega \end{bmatrix} = \begin{bmatrix} \dot{\alpha} + \dot{\gamma}\,\sin\beta \\ \dot{\beta}\,\cos\beta - \dot{\gamma}\,\cos\beta\,\sin\alpha \\ \dot{\gamma}\,\cos\alpha\,\cos\beta + \dot{\beta}\,\sin\alpha \end{bmatrix}. \qquad (2.1)$$

The equations of motion follow from the theorem of impulse and the theorem of angular momentum for the center of mass S. In the inertial system the equations are :

$$m\,\ddot{y}_1 = G_i + P_i\,,$$

$$I_{ij}\,\dot{\Omega}_j + \varepsilon_{ijk}\,\Omega_j\,I_{k\ell}\,\Omega_\ell = M_i\,, \qquad (2.2)$$

where $G_i = \begin{bmatrix} 0, 0, -mg \end{bmatrix}^T$ is the rotor weight, P_i and M_i are restoring forces and moments that are acting from the elastic shaft on the rotor. They follow from the linear Maxwell relations for deflection and angle

$$y_1 + e_1 = -\left[a_0 P_1 + c_0 M_2\right], \quad y_2 + e_2 = -\left[a_0 P_2 - c_0 M_1\right]$$

(2.3)

$$\beta = -\left[c_0 P_1 + b_0 M_2\right], \quad \alpha = -\left[c_0 P_2 + b_0 M_1\right].$$

The influence numbers are

(2.4) $a_0 = \ell^3 / 3EI, \quad b_0 = \ell / EI, \quad c_0 = \ell^2 / 2EI,$

with the length ℓ of the elastic shaft and the stiffness EI and for e_i, the transformation $e_i = c_{ij} \bar{\bar{e}}_j$ is valid.

From (2.3)

$$P_i = -\begin{bmatrix} b(y_1 + e_1) - c\beta \\ b(y_2 + e_2) + c\alpha \\ -mg \end{bmatrix}, \quad M_i = -\begin{bmatrix} a\alpha + c(\dot{y}_2 + e_2) \\ a\beta - c(\dot{y}_1 + e_1) \\ -M \end{bmatrix}$$

(2.5)

is obtained with the abbreviations

(2.6) $a = a_0 / (a_0 b_0 - c_0^2), \quad b = b_0 a / a_0, \quad c = c_0 a / a_0$

and M the driving torque.

The nonlinear equations of motion can now be derived from (2.2). The equations, however, come out to be so complicated that calculation or a good understanding of the physical meaning is difficult. Therefore the equations are linearized

1. with respect to the variables $y_i, \alpha, \beta,$
   and their time-derivatives,

2. with respect to the eccentricity e .

The first assumption means that motions in the neighbourhood of the undisturbed rotation on the vertical are investigated. The validity of this assumption has to be checked by means of the calculated solutions. The second assumption implies that the eccentricity, being an undesired inaccuracy of manufacturing, shall naturally be small. Furthermore, the driving torque is supposed to be $M = 0$ and the rotor shall turn with constant angular velocity $\dot{\gamma} = \Omega$. This rotor velocity is then independent of the other rotor motions, but of course only so long as they are sufficiently small. The coupling of the rotor motions to the non-stationary rotor velocity is shown in Chapter 6. 1.

Hence it follows

$$m \ddot{y}_1 + b y_1 - c \beta = e b \cos \Omega t$$

$$m \ddot{y}_2 + b y_2 + c \alpha = e b \sin \Omega t$$

$$A \ddot{\alpha} + C \Omega \dot{\beta} + a \alpha + c y_2 = e (c - mg) \sin \Omega t$$

$$(2.7)$$

$$A \ddot{\beta} - C \Omega \dot{\alpha} + a \beta - c y_1 = -e (c - mg) \cos \Omega t .$$

A physical interpretation of these equations of motion is facilitated by regarding the deflection of the center W of the rotor :

Substituting $y_i = u_i - c_{ij} \, \bar{\bar{e}}_j$ into (2.7) leads to

$$m\ddot{u}_1 + b u_1 - c\beta = e m \Omega^2 \cos \Omega t$$

$$m\ddot{u}_2 + b u_2 + c\alpha = e m \Omega^2 \sin \Omega t$$

(2.8)

$$A\ddot{\alpha} + C\Omega\dot{\beta} + a\alpha + c u_2 = -e mg \sin \Omega t$$

$$A\ddot{\beta} - C\Omega\dot{\alpha} + a\beta - c u_1 = e mg \cos \Omega t.$$

The physical significance of the terms on the left side is obvious as they are inertia forces, restoring forces and moments caused by deflection and inclination and gyroscopic moments as well as centrifugal forces and gravity moments on the right side. It is a well-known method to establish the equations of motion by balancing these forces and moments and thus reducing the task to a static problem [3]. One has to be careful, however, in choosing a suitable coordinate system, in determining the gyroscopic moment and in not anticipating a rotor whirling that actually should be calculated by this statement [2].

The equations of the transverse motion of the rotor can be clearly arranged in matrix form. The matrix notation is particularly useful for the presentation and calculation of higher-order systems but principal handling and advantages become already apparent by this simple example. The variables form the vector

(2.9) $$v = \left[y_1, \beta, y_2, -\alpha \right]^T.$$

Then equ. (2.7) reduces to the matrix equation

$$M\ddot{v} + G\dot{v} + Fv = f \qquad (2.10)$$

with the matrix of masses

$$M = \text{diag}\left[m, A, m, A\right],$$

the skew-symmetric gyroscopic matrix

$$G = \begin{bmatrix} 0 & 0 & 0 & 0 \\ 0 & 0 & 0 & C\Omega \\ 0 & 0 & 0 & 0 \\ 0 & -C\Omega & 0 & 0 \end{bmatrix},$$

the symmetric matrix of the restoring coefficients

$$F = \begin{bmatrix} b & -c & 0 & 0 \\ -c & a & 0 & 0 \\ 0 & 0 & b & -c \\ 0 & 0 & -c & a \end{bmatrix},$$

and the vector of the disturbing unbalance

$$f = \left[e\, b\, \cos\Omega t, \ -e\,(c - mg)\cos\Omega t, \right.$$

$$\left. e\, b\, \sin\Omega t, \ -e\,(c - mg)\sin\Omega t\right]^{T}.$$

A still further compactness of (2.10) is achieved by introducing complex variables

(2.11)
$$y = y_1 + i y_2$$
$$\delta = \beta - i \alpha .$$

These variables characterize the radial symmetry of the structure. They reduce (2.10) to

(2.12)
$$\bar{M} \ddot{w} - i \bar{G} \dot{w} + \bar{F} w = \bar{r} e^{i \Omega t},$$

where

$$w = [y, \delta]^T,$$
$$\bar{M} = \text{diag} [m, A],$$
$$\bar{G} = \text{diag} [0, C\Omega],$$

$$F = \begin{bmatrix} b & -c \\ -c & a \end{bmatrix},$$

$$\bar{r} = [e b, -e(c - m g)]^T.$$

2.2. Stability, forward and reverse whirl.

Stability of the system (2.10) can be judged already by a general theorem of Thomson-Tait. An extensive survey on matrix stability theorems is given by MAGNUS [4, 5]. The conservative system (2.10) is stable, if the restoring matrix F is positive definite or if

$$ab - c^2 > 0$$

is valid, i.e. if the support of the rotor is statically stable.
Then the gyroscopic matrix G cannot remove stability no mat-
ter how fast the rotor turns or if the rotor has an elongated
form $(C < A)$ or a flat one $(C > A)$.

    The motion will in one form or another be subject to
damping. Let, for example, the disk vibrate in a damping fluid,
supposing that the damping force is proportional to the rate of
deflection then (2.10) has to be completed by a damping matrix
D   and

$$M\ddot{v} + (G + D)\dot{v} + Fv = f, \qquad (2.14)$$

where

$$D = \begin{bmatrix} d & 0 & 0 & 0 \\ 0 & d_1 & 0 & 0 \\ 0 & 0 & d & 0 \\ 0 & 0 & 0 & d_1 \end{bmatrix}.$$

    As D is positive definite the aforesaid stability criteri-
on, that was only sufficient, now becomes necessary as well. A
stable solution of (2.10) will now be asymptotically stable. This
special kind of damping is named external damping, as the
damping forces react on the externally fixed parts of the sup-

port. By way of contrast an internal damping, resulting from internal friction in the material of the shaft, cannot improve the system's behavior but can even cause instability as shown in chapter 6.2.

Now the eigenvalues of the system (2.12) are determined. The function

$$w = W e^{\lambda t}$$

satisfies the homogeneous part of (2.12) if the characteristic equation of λ holds

$$\det \left[\bar{M} \lambda^2 - i \bar{G} \lambda + \bar{F} \right] = 0 .$$

In detail, with $\lambda = i\omega$, it follows

(2.15) $m A \omega^4 - m C \Omega \omega^3 - \omega^2 (ma + Ab) + bC\Omega\omega + ab - c^2 = 0 .$

This equation has four real roots that are plotted in Fig. 3 against rotor velocity as dimensionless variables. It is seen that there are two positive natural frequencies ω_2, ω_4 and two negative ones ω_1, ω_3 that are all different except for zero rotation where $\omega_1 = -\omega_2$ and $\omega_3 = -\omega_4$. The curves are symmetrical which means that for $+\Omega$ and $-\Omega$ the same values occur, in other words, the four natural frequencies do not care whether the disk rotates clockwise or counterclockwise. When $\Omega \to \infty$ the natural frequencies tend asymptotically to constant values

$$\omega_1 \to 0, \quad \omega_2 \to 2r, \quad \omega_3 \to -2r, \quad \omega_4 \to (C/A)\Omega$$

as can be seen by substituting in (2.15). Hence, r is the frequency of the flexural vibrations when instead of the rotor a concentrated mass is mounted to the end of the shaft. Then,

$$r^2 = \frac{ab - c^2}{am} = \frac{3\,EI}{\ell^3 m}$$

holds. The fast natural frequency is called the nutation frequency, the lowest one is the precession frequency. The solution of the homogeneous equation (2.12) is given by

$$w_i = \sum_j W_{ij}\, e^{\lambda_j t} = \sum_j W_{ij}(\cos \omega_j t + i \sin \omega_j t),$$
$$i = 1, 2,$$
$$j = 1, \dots 4.$$
(2.16)

It is seen that for positive natural frequencies ω_2, ω_4 there are natural modes describing a circular motion of the center of mass in positive direction, i. e. in the same sense as the rotor angular velocity Ω (forward whirl). For negative natural frequencies ω_1, ω_3 the motion occurs as a reverse whirl in a sense opposite to Ω. Natural modes of the rotor can be excited by suitable disturbing forces, for example by an impulse. The complex amplitudes W_{ij} are predetermined by the initial condition. The normal modes are superimposed and form

the resulting motion of the disk. It is easily understood that for
any rotor velocity, forward whirl as well as reverse whirl can
occur. Of course, the whirling frequency in general will not be
equal to the rotor angular frequency Ω. Both motions will decay according to damping if they are not sustained by suitable
disturbances.

2. 3. Critical speed and resonance.

When the normal modes are excited in the   cycle of a
natural frequency, the amplitudes of the undamped vibrations
grow unboundedly. Practically these resonance amplitudes are
bounded to a finite value by damping or by nonlinearities. As
the rotor will always be slightly unbalanced, there will always
act an excitation on the rotor with the frequency and in the sense
of the rotor velocity. Resonance occurs when the rotor velocity
Ω coincides with a natural frequency. From Fig. 3 (see
next page) it follows that these resonance frequencies or critical speeds are given by the intersection points of the straight
line $\omega_i = \Omega$ with the frequency curves. The critical
speeds can be calculated as well from (2.15) with $\omega_i = \Omega$.

The term critical speed is often used in an inexact
way for the frequencies that result from the intersection of the
straight line $\omega_i = - \Omega$ with the frequency curves ω_1 and ω_3.
However, the natural modes (reverse wheel) corresponding to
these natural frequencies are as stated previously not excited

and do not lead to any resonance.

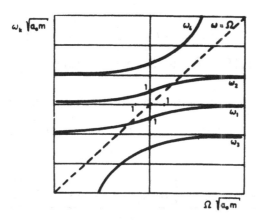

Fig. 3

The vibrations forced by rotor unbalance are calculated explicitly. The statement

$$w = W e^{i \Omega t}$$

is substituted into (2.12) and leads to

$$\left[- \bar{M} \Omega^2 + i \bar{G} \Omega + \bar{F} \right] W e^{i \Omega t} = \bar{r} e^{i \Omega t} ,$$

and thus

$$W = \left[- \bar{M} \Omega^2 + i \bar{G} \Omega + \bar{F} \right]^{-1} \bar{r} ,$$

where W is called the complex transfer function. In detail

$$W_1 = \frac{e\,b\,(C-A)\Omega^2 + e\,b\,a - c\,e\,(c-mg)}{-\Omega^4\,m\,(C-A) + \Omega^2\left[b\,(C-A) - ma\right] + ab - c^2},$$

$$W_2 = \frac{e\,b\,c - e\,(c-mg)\,(-m^2+b)}{-\Omega^4\,m\,(C-A) + \Omega^2\left[b\,(C-A) - ma\right] + ab - c^2}.$$

Resonance occurs when the denominator is zero (and the numerator does not vanish at the same time), i.e. for the critical speeds

$$\Omega^2_{1/2} = \frac{b(C-A) - ma \pm \sqrt{\left[b(C-A)-ma\right]^2 + 4(ab-c^2)(C-A)m}}{2\,m(C-A)}.$$

(2.17)

Only positive values for Ω^2 will give a real result for Ω. From (2.17) it follows that two critical speeds come out for an elongated rotor ($C < A$) whereas only one critical speed can exist for a flat rotor ($C > A$). The same result is seen from Fig. 3. The curve of the nutation frequency tends to the asymptote $\omega_4 = (C/A)\Omega$. For a flat gyro with $C/A > 1$ no intersection with the straight line $\omega_j = \Omega$ is possible, but only for the elongated gyro with $C/A < 1$.

It shall be mentioned that there are still other sources of disturbing forces as, for example, the weight of the rotor when the rotor is supported horizontally [2], or external periodic forces. They all can cause resonance phenomena sometimes characterized by the term secondary resonance.

3. System of symmetrical high-speed gyros

3.1. Fundamental equations of motion.

The equations of motion for gyroscopic systems are derived by MAGNUS [4] by applying Lagrange's equation and the function of Routh. In the following, the assumptions that underlie the equations, as they are used here, are made up and motivated :

1. Let the rotors be symmetrical with respect to their axes of rotation and let the moments about these axes be balanced. Then, with the kinetic energy T and the generalized forces Φ_k that act on the motion of the coordinates Φ_k about the axes of rotation, it follows

$$\frac{\partial T}{\partial \Phi_k} = 0 \; , \qquad Q_k = 0$$

and hence

$$\frac{\partial T}{\partial \dot{\Phi}_k} = \text{const} = H_k \, .$$

This means that the coordinate Φ_k describing the rotation about the axis k is a cyclical coordinate and that the moment of momentum H_k of the rotor is constant.

2. Let the remaining non-cyclical coordinates be small

with respect to their deviations from given values. Then the
equations of motion can be linearized by the method of small
vibrations.

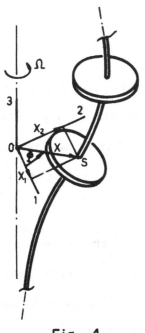

Fig. 4

3. Let the variables
of the system be such that they
can be arranged as complex
variables. This requires a ra
dial symmetry of the physical
model, a property characterized as well by cyclical coordi
nates. Thus, a complex variable shows for example the ra
dial displacement x of a rotor
denoting amount and direction
(Fig. 4):

$$x = x_1 + i x_2 = X e^{i\Phi},$$

(3.1)

$$X = \sqrt{x_1^2 + x_2^2} \qquad \tan \Phi = x_2 / x_1$$

4. Let the rotors be turning rapidly; i.e., their moment of momentum and thus their kinetic energy is very much greater than the kinetic energy of the other vibrating parts of the system (bearings, supports).

The assumptions 1 and 2 enable the evaluation of a linear system of differential equations with constant coefficients. The matrix form reads

$$M \ddot{v} + (D + G) \dot{v} + F v = 0 , \qquad (3.2)$$

where

$$v = [v_i]^T, \qquad i = 1 ,\ldots m ,$$

is the vector of displacement with m coordinates according to the m degrees of freedom of the system, $M = [m_{ij}]$ is a symmetrical, positive definite mass-matrix, $D = [d_{ij}]$ is a symmetrical damping matrix, $G = [g_{ij}]$ is a skew-symmetrical gyroscopic matrix. The elements of the gyroscopic matrix depend linearly on the moment of momentum of a rotor. Thus it can be written

$$g_{ij} = H p_{ij} ,$$

where H is a parameter denoting the largeness of the moment of momentum and proportional to the rotor velocity. The symmetrical matrix $F = [f_{ij}]$ characterizes conservative forces. Non-conservative forces that can positively occur are not considered here.

Because of the third assumption the number of degrees of freedom is even (m = 2n) and the matrices show further properties of symmetry. They can be separated into 4 quadratic submatrices A_{ab} (a,b = 1,2), respectively. For symmetrical matrices $A_{11} = A_{22}$ and $A_{12} = A_{21} = 0$, for the skew-symmetrical gyroscopic matrix $A_{11} = A_{22} = 0$ and $A_{12} = -A_{21}$ where only the main and the minor diagonals of the submatrices are (partly) filled up. Then, the complex variables

$$w_k = v_k + i\,v_{n+k}$$

build the vector

$$w = \left[w_k \right]^T, \quad k = 1,\dots,n$$

and the system (3.2) reduces from the order 2m to the order m having the form

(3.3) $$\bar{M}\,\ddot{w} + (\bar{D} - i\bar{G})\,\dot{w} + \bar{F}\,w = 0,$$

where

$$\bar{M} = \left[m_{k\ell} \right] \qquad k,\ell = 1,\dots,n$$
$$\bar{D} = \left[d_{k\ell} \right]$$
$$\bar{G} = \left[g_{k,\,n+\ell} \right]$$
$$F = \left[f_{k\ell} \right],$$

The completion of this equation of motion by terms on the right side of the equation caused by disturbing forces is dealt with in Chapt. 3. 3.

3. 2. Stability and eigenvalues.

The stability of the solutions of (3. 2) can be largely determined on account of the structure of the matrices. Among the numerous stability criteria [4] the one of Thomson-Tait says : The statically stable equilibrium of the system remains stable, even if any gyroscopic and dissipative forces are added. The equilibrium is statically stable, if the matrix F is positive definite. Dissipative forces are forces that damp the motion. Furthermore, the motion is asymptotically stable, if the damping is positive definite or if it is positive semidefinite and the damping acts pervasively on the system [6]. The theorem is not only sufficient but also necessary as a statically unstable position of equilibrium cannot be stabilized by gyroscopic forces, if at the same time there are also dissipative forces. In the following it is assumed that the solutions are stable.

Now some statements concerning the eigenvalues are derived from (3. 3.). The relationship

$$w = We^{\lambda t},$$

substituted into (3. 3) leads to the characteristic equation

(3.4) $$\det\left(\bar{M}\lambda^2 + \bar{D}\lambda - i\,\bar{G}\lambda + \bar{F}\right) = 0.$$

The system without any gyroscopic effects, $\bar{G} \equiv 0$, that is regarded at first, is a damped vibrational system with real coefficients, the eigenvalues of which are conjugate complex

$$\lambda_k = \delta_k \pm i\,\omega_k \qquad k = 1,\ldots,n.$$

The motion of the system consists of superimposed natural modes of the form

(3.5) $$w_k = W_k\, e^{\delta_k t}\, e^{\pm i\,\omega_k t}.$$

With regard to the physical meaning of the complex representation (3.1), the ensuing motion of (3.5) is a spiral. Its radius decreases by $e^{\delta_k t}$ (δ_k is negative, as the system is supposed to be stable), the sense of whirling is given by the sign of ω_k . The spiral whirls about the nominal axis of the rotor rotation in the same sense (forward) or in the opposite sense (reverse) to it. There are as many forward natural modes as reverse ones with the same frequencies, respectively.

It can be expected that this principal behavior will not change by adding gyroscopic forces. There will be two groups of natural modes of the same size, even when the frequency of the forward and the reverse whirl will differ on account of the increasing gyroscopic effect.

Some statements on the influence of gyroscopic forces on the eigenvalues follow from (3.4). Written as a polynomial it reads

$$\Delta(\lambda) = b_m \lambda^m + b_{m-1} \lambda^{m-1} + \ldots + b_1 \lambda + b_0 = 0. \qquad (3.6)$$

As $m = 2n$ this polynomial contains the term $b_n \lambda^n$. The factor b_n depends on the elements $g_{k\ell}$ of the gyroscopic matrix [4]. As these elements depend linearly on the moment of momentum H the factor b_n can be developed into series of H

$$b_n(H) = g_n(-i)^n H^n + \ldots$$

beginning with $g_n = \det[p_{k\ell}]$. The adjoining terms of b_n are functions of H as well, but with exponents that decrease with the distance from the middle term b_n. The first and the last coefficient are independent of H as $b_m = \det \bar{M}$ and $b_0 = \det \bar{F}$ holds. The polynomial can be separated into two parts with increasing and decreasing exponents of H [7]:

$$\Delta_1 = b_{2n} \lambda^{2n} + b_{2n-1} \lambda^{2n-1} + \ldots + b_n \lambda^n,$$

$$\Delta_2 = b_n \lambda^n + \ldots + b_1 \lambda + b_0.$$

In order to know the tendency of the dependence of the eigenvalues on H each of these expressions Δ_1 and Δ_2 is equalled to zero. From the first one come out roots the

amounts of which are proportional to H, for the second one
they decrease with $1/H$. The roots of the first group characterize fast nutational motions, the roots of the second group
belong to the slow precessions. A complete separation of the
roots into these two groups is only possible for $\det \bar{G} \neq 0$.
In the case of a singular gyroscopic matrix the adjoining coefficients to b_n (their number depends on the defect of the matrix
$\bar{G}$) are of the same order with respect to H. Therefore, in
the extreme case $H \to \infty$ there exist some roots of the characteristic equation that tend to a constant value, independent
on H. The motions governed by them are called pendulous
motions.

For very fast gyroscopes the terms with the highest
orders in H dominate. Therefore the roots can approximate-
ly be determined by equalling two neighboring terms, respectively, of the characteristic polynomial. The highest nutation
frequency comes out, for example, as $\lambda_m = b_{m-1}/b_m$.
In order to know the influence of the gyroscopic effects on the
damping coefficient and the natural frequency of the most characteristic eigenvalues this method is continued in detail for
the highest and the lowest terms of (3.6). As shown their coefficients can be developed in series of

$$b_m \lambda^m + (iHg_{m-1} + a_{m-1})\lambda^{m-1} + (-H^2 h_{m-2} + iHg_{m-2} + a_{m-2})\lambda^{m-2} +$$

(3.7)
$$+ \ldots + (-iHg_1 + a_1)\lambda + b_0 = 0.$$

Calculating at first the highest root $\lambda_N = \delta_N + i\omega_N$, where $\omega_N = \mu H$ is proportional to H as it was made reasonable just before, the relationship

$$\lambda^m = (i\mu H)^m \left[1 - i \binom{m}{1} \frac{\delta_N}{\mu H} - \binom{m}{2} \left(\frac{\delta_N}{\mu H} \right)^2 + \dots \right]$$

is used for developing (3.7) and hence

$$b_m (i\mu H)^m \left[1 - im \frac{\delta_N}{\mu H} - \dots \right] +$$

$$+ (iHg_{m-1} + a_{m-1})(i\mu H)^{m-1} \left[1 - i(m-1) \frac{\delta_N}{\mu H} - \dots \right] +$$

$$+ (-H^2 h_{m-2} + iHg_{m-2} + a_{m-2})(i\mu H)^{m-2} \left[1 - \dots \right] + \dots = 0 .$$

Rearranged with respect to falling orders of H

$$H^m \left[b_m (i\mu)^m + i g_{m-1} (i\mu)^{m-1} - h_{m-2} (i\mu)^{m-2} + \dots \right] +$$

$$+ H^{m-1} \left[b_m m \delta_N (i\mu)^{m-1} + a_{m-1} (i\mu)^{m-1} + i g_{m-1} (m-1) \delta_N (i\mu)^{m-2} + \right.$$

$$\left. + i g_{m-2} (i\mu)^{m-2} \right] + \dots = 0 . \tag{3.8}$$

For $H \to \infty$ the first bracket has to be zero in order to satisfy the equation. From this condition the proportionality factor μ has to be evaluated. For the sake of simplicity, it is assumed that the gyroscopic system contains only one rotor so that $h_{m-2} = 0$. Thus,

$$(3.9) \qquad\qquad \mu = -g_{m-1}/b_m$$

holds. The highest frequency of nutation is then $\omega_N = \mu H$, the sense of whirling is determined by the sign of g_{m-1} as always $b_m > 0$. If the gyroscopic system contains more rotors further nutational frequencies exist.

The calculation of the damping coefficient δ_N needs zeroing of the bracket for H^{m-1} :

$$b_m \, m \, \delta_N + a_{m-1} + g_{m-1}(m-1)\delta_N/\mu + g_{m-2}/\mu = 0.$$

Hence, with (3.9)

$$(3.10) \qquad\qquad \delta_N = -\frac{a_{m-1}}{b_m} + \frac{g_{m-2}}{g_{m-1}} \ .$$

In general the nutation will be damped. It is quite possible, however, that for special structure of the gyroscopic system

$$(3.11) \qquad\qquad g_{m-1} \, a_{m-1} = b_m \, g_{m-2}$$

holds, i. e. the nutation will be undamped for high rotor velocities.

   For the lowest eigenvalue $\lambda p = \delta_p + i\omega_p$ the last
two terms of (3.7) are dominant and lead to

$$\lambda_p = \frac{b_0 a_1 + i b_0 g_1 H}{a_1^2 + g_1^2 H^2} .$$

   For $H \to \infty$ both the damping $\delta_p \sim 1/H^2$ and the precession frequency $\omega_p \sim 1/H$ tend to zero.

   In Chapter 4 a gyroscopic system is presented the
technical performance of which is decisively governed by its
damping behavior for high rotor velocities.

3. 3. Critical speed and resonance.

   Disturbing forces that act on the system are caused
first of all by small rotor unbalances and lead in general to a
disturbance vector $f(t)$ on the right side of the equation of
motion (3.2). The behavior is, in the last analysis, already
given by the solutions of the homogeneous system (Chapt. 3. 2),
but nevertheless it shall be discussed briefly.

   Disturbances by rotor unbalances are caused by the
rotation $\dot{\Phi}_k = \Omega_k$ of a slightly unsymmetrical rotor. Let
the unsymmetry, however, be so small that the variable Φ_k
remains as assumed a cyclical variable, when the equations
are linearized. The unbalance forces can form, quite similar

to the deviation of Fig. 4, a complex quantity.

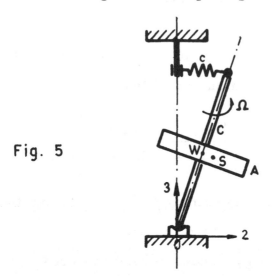

Fig. 5

For the sake of simplicity it is assumed that the angular velocities of the rotors are all alike $\dot{\Phi}_k = \Omega$ and thus the vector
of disturbances of (3. 3) is $\bar{r}\, e^{i\Omega t}$. Then, the particular solution

(3. 13) $w = W\, e^{i\Omega t}$

satisfies the nonhomogeneous equation

(3. 14) $\bar{M}\,\ddot{w} + (\bar{D} - i\,\bar{G})\dot{w} + \bar{F} w = \bar{r}\, e^{i\Omega t}.$

The frequency and the sense of whirling of the forced
motion (3. 13) corresponds to that of the excitation. This implies
that excitations by the angular rotor velocity can cause only
forward whirling and no reverse whirling. Substitution of
(3. 13) into (3. 14) leads to

$$\left[-\Omega^2\bar{M} + (i\bar{D} + \bar{G})\Omega + \bar{F}\right]W e^{i\Omega t} = \bar{r} e^{i\Omega t}$$

and by denoting the brackets $[..] = \Delta$ the complex resonance function

$$W = \Delta^{-1}\bar{r} = \frac{adj\,\Delta}{det\,\Delta}\,\bar{r} \qquad (3.15)$$

and the particular solutions

$$w = \Delta^{-1}\bar{r}\,e^{i\Omega t} = \frac{adj\,\Delta}{det\,\Delta}\,\bar{r}\,e^{i\Omega t} \qquad (3.16)$$

come out, provided that the inverse Δ^{-1} exists. The inverse does not exist, if Δ is singular, i.e. if $det\Delta = 0$. This can only occur if the system is undamped and the exciting frequency Ω is equal to a natural frequency ω_i. Then, the particular solution increases unboundedly in the case of resonance (the statement (3.13) is then not valid). In the case of apparent resonance it takes finite values as shown below.

Resonance occurs when the resonance function (3.15) as a function of the exciting frequency Ω becomes a maximum. This is the case when the denominator comes to a minimum unless it is cancelled by a minimum of the numerator reached at the same time. This effect is called apparent resonance. Physically resonance means that a natural frequency ω_i is equal to the exciting rotation frequency Ω of the rotor, the so-called critical speed and furthermore that the exciting forces do really excite the natural modes. Apparent resonance occurs if the

last condition is not fulfilled. For example, resonance caused
by rotor unbalance can only occur as forward whirl not as reverse whirl even if the magnitude of the natural frequency is
equal to the rotor frequency [8, 9, 10].

The term "apparent resonance" shall be specified by
a simple example [11] : A rotor is standing on the bottom of
its axis and its upper end is elastically strapped down (Fig. 5).
Let its motion, characterized by the angular rotor velocity
and the cardanic angles (Fig. 2) α, β be described by

$$A\ddot{\beta} - C\Omega\dot{\alpha} + c\beta = k \cos \Omega t$$

(3.17)

$$A\ddot{\alpha} + C\Omega\dot{\beta} + c\alpha = -k \sin \Omega t.$$

A solution is assumed

$$\alpha = \alpha_1 \cos \Omega t + \alpha_2 \sin \Omega t$$

(3.17)

$$\beta = \beta_1 \cos \Omega t + \beta_2 \sin \Omega t$$

and its substitution into (3.17) leads to

$$\alpha_1 = \beta_2 = 0$$

(3.19)

$$\beta_1 = -\alpha_2 = -k \; \frac{(A+C)\Omega^2 - c}{\left[(A-C)\Omega^2 - c\right]\left[(A+C)\Omega^2 - c\right]}$$

The denominator of the resonance function (3.19) has
four roots Ω_i, $(i=1,...,4)$, corresponding to the number of nat-

ural frequencies of the system. At the same time the numerator shows two roots $\Omega_{1,2}$ that cancel two roots of the denominator so that indeed there remain only two resonance frequencies or critical speeds at

$$\Omega_{3/4} = \pm \sqrt{c/(A-C)} \ .$$

Only the natural modes of the forward whirl are excited here.

The hidden resonance at

$$\Omega_{1/2} = \pm \sqrt{c/(A+C)}$$

is called apparent resonance as the amplitudes remain finite
here. The normal modes of the reverse whirl are not excited
and do not lead to resonance.

One conclusion following from it is that it does not in
general suffice to study the denominator of a resonance function in order to obtain from its roots resonance frequencies-
one has to consider the numerator as well. On the other side
it is not to be expected that a special excitation of a system, as
it is given by rotor unbalances, will cause resonance at all
natural frequencies in order to determine eigenvalues by the
aid of resonance curves. Sometimes in measuring techniques
this eventually is disregarded and even with studying the be-

havior of systems by the method of transfer matrices this effect can cause difficulties.

4. Rotor with elastically supported bearings

4.1. Gyroscopic motion.

The general results of Chapt. 3 are now applied to the investigation of an elastically supported gyro [11] Fig. 6 shows the theoretical structure of the gyro.

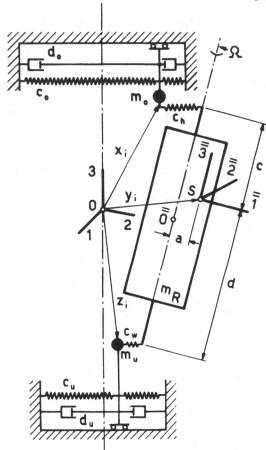

The nearly symmetrical rotor is elastically centered to the vertical by two bearings that are elastically suspended themselves. Their motion is linearly damped. The rotor is driven about its axis of symmetry and is expected to reach a high angular velocity. The investigation has to show whether and how much the system suits to this aim. At first the equations of motion for this system

Fig. 6

with eight degrees of freedom will be established. There are

six translational degrees of freedom, two for the upper and the
lower bearing and the center of mass of the rotor, respectively,
the rotor has two rotational degrees of freedom, the angular
velocity about its axis of symmetry be a constant. For a symmetrical gyro the rotation about the vertical is a particular
solution. It has to be asymptotically stable. For the nearly
symmetrical gyro the deviations of its axis of symmetry from
the vertical shall be small. Therefore the highly nonlinear e-
quations of motion can be linearized provided the variables are
chosen suitably. According to Fig. 6, two coordinate systems
are used. The transformation of the inertially fixed 1 2 3 - sys-
tem into the body-fixed $\bar{1}\ \bar{2}\ \bar{3}$ - system is described as in Chapt.
2 by the cardanic angles α, β, γ (Fig. 2) where $\alpha, \beta \ll 1$ denote
the inclination of the rotor with respect to the vertical and $\dot{\gamma} =$
$= \Omega$ is the angular rotor velocity about its axis of symmetry.
In addition to the self-explaining symbols of Fig. 6 there are
moments of inertia, defined by the tensor of inertia for the cen-
ter of mass of the rotor related to the $\bar{1}\ \bar{2}\ \bar{3}$ - coordinate system

$$\bar{I}_{ij} = \begin{bmatrix} A & -F & -E \\ -F & A & -D \\ -E & -D & C \end{bmatrix}$$

with D, E, F as small quantities, and the vertical forces N, K
between the rotor and the upper or lower bearing, respectively.

The forces are positive when they act on the rotor in the direction of the 3-axis.

The theorem of moment of momentum is now establish̲ed for the rotor and the theorem of impulse is applied to the motion of the upper and lower bearing and the center of mass of the rotor in a similar way as it was done in Chapt. 2 for the rotor on overhung shaft. Hence, in inertially fixed coordinates the equations of motion show the well-known form (Chapt. 3.1)

$$M \ddot{v} + (D + G)\dot{v} + F v = f(t) . \qquad (4.1)$$

The coefficient matrices are 8 x 8 matrices and the vector is composed of the variables

$$v = \left[x_1, y_1, z_1, \beta ; x_2, y_2, z_2, -\alpha \right]^T .$$

The transverse displacements in the direction of the 1- and 2- axis are arranged in a way that they can be easily transformed into complex quantities. By $x = x_1 + i x_2$, $y = y_1 + i y_2$, $z = z_1 + i z_2$, $\Theta = \beta - i\alpha$ the complex vector has the form

$$w = \left[x, y, z, \Theta \right]^T , \qquad (4.2)$$

the disturbance vector is

$$\bar{f} = \bar{r} e^{i\Omega t}$$

and the equ. (4.1) changes into

(4.3) $\bar{M}\ddot{w} + (\bar{D} - i\bar{G})\dot{w} + \bar{F}w = \bar{r} e^{i\Omega t}$,

where the coefficient matrices are 4 x 4 submatrices of the coefficient matrices of (4.1) :

$$\bar{M} = \text{diag} \left[m_0, m_R, m_u, A \right],$$

$$\bar{D} = \text{diag} \left[d_0, 0, d_u, 0 \right],$$

$$\bar{G} = \text{diag} \left[0, 0, 0, C\Omega \right],$$

$$\bar{F} = \begin{bmatrix} c_0 + c_h & -c_h & 0 & -c_h c \\ -c_h & c_h + c_w & -c_w & c_h c - c_w d \\ 0 & -c_w & c_w + c_u & c_w d \\ -c_h c & c_h c - c_w d & c_w d & c_h c^2 + c_w d^2 + Nc - Kd \end{bmatrix},$$

$$\bar{r} = \left[-c_h a, (c_h + c_w) a, -c_w a, (c_h c - c_w d) a + \right.$$
$$\left. + (N + K) a + (E + iD)\Omega^2 \right]^T.$$

The disturbance $\bar{r}\, e^{i\Omega t}$ is caused by static and dynamic unbalance of the rotor, i. e., by an eccentricity **a** of its center of mass and by products of inertia E and D.

Stability can be judged by the theorem of Thomson-Tait (Chapt. 3.2). Necessary and sufficient condition for stability is that F is positive definite, i.e. that the system is statically stable. This is the case if

$$(Nc - Kd)\left[(c_0 + c_h)\, c_u c_w + (c_u + c_w)\, c_0 c_h\right] +$$

$$+.\ c_0 c_h c_u c_w (c + d)^2 > 0 \qquad (4.4)$$

holds. As the damping matrix is only positive semidefinite asymptotical stability on the basis only of the damping matrix is not secured but it can be supposed. Indeed the damping is pervasive as can be shown by calculating the eigenvalues $\lambda_k = \delta_k + i\, \omega_k$ for a numerical example. It turns out that the real parts δ_k are all negative and thus the solutions are asymptotically stable. The natural frequencies ω_k are shown in Fig. 7 (see next page) depending on the rotor frequency Ω. Their general behavior could already have been predicted by the aid of the general theorems for gyroscopic systems (Chapt. 3.2). One of the natural frequencies, the nutation frequency ω_8 increases with the rotor frequency Ω. Another one, the precession frequency ω_1 tends to zero. The middle ones, the pendulous frequencies tend to constant values. Both the gyroscopic frequencies make trouble as their real parts δ_1, δ_8 tend to zero when

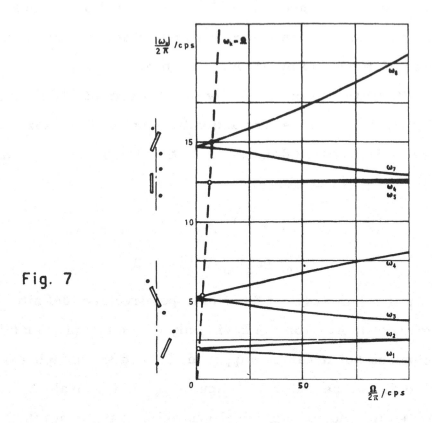

Fig. 7

the rotor velocity increases. For the precession this had to be
expected on account of (3. 12) but as to the nutation this undesired effect occurs because the parameter tuning (3. 11) is fulfilled on account of the special supporting structure of the rotor.

On the left side of Fig. 7 the natural modes are outlined in a constellation that would be seen by an observer rotating with Ω (the point stands for the bearing, the dash for the
rotor). For the lowest pair of natural frequencies bearings and
rotor vibrate in phase. For the highest natural frequencies,
bearings and rotor are out of phase. Of course, the motions go

only then so distinctly in one plane as outlined, if there is no or low damping. Damping has the effect that - seen from the rotat ing observer - the motion does not occur in a plane but spatial- ly, the axis of symmetry does not turn any more in a cone but on a hyperboloid about the vertical. The four lower branches $\omega_1, \omega_3, \omega_5, \omega_7$ of the pairs of frequencies belong to natural modes that whirl opposite to the rotation of the rotor. The four upper branches belong to forward motion in the sense of the rotation of the rotor. For the calculation of the natural frequencies and modes the following hint may be useful. The displacements $\mathbf{w}$ and their time- derivatives $\dot{\mathbf{w}}$ form the state space vector

$$ \mathbf{s} = \begin{bmatrix} \mathbf{w} \\ \dot{\mathbf{w}} \end{bmatrix} $$

and with it the homogeneous part of (4. 3) transforms into the state space equation

$$ \dot{\mathbf{s}} = R\,\mathbf{s}, $$

where

$$ R = \left[\begin{array}{c|c} 0 & I_4 \\ \hline -\bar{M}^{-1}(\bar{D} - i\,\bar{G}) & -\bar{M}^{-1}\bar{F} \end{array} \right] $$

is an 8 x 8 matrix. The solution of

(4. 5) $\left[\lambda I_8 - R\right] s = 0$

leads to the eigenvalues

$$\lambda_k = \delta_k + i\omega_k \qquad k = 1,\ldots,8$$

and the corresponding eigenvectors

$$\overset{k}{s} = \overset{k}{s}{}' + i\,\overset{k}{s}{}''.$$

Standard programs for solving (4. 5) are available at computing centers.

The investigation of the homogeneous equation of motion informed about stability and free motion. Now the particular solution of (4. 3), is considered. It follows in detail from (3. 15) and (3. 16) as

$$w = W\,e^{i\Omega t}.$$

The complex resonance function W characterizes amplitude and phase of the forced motion. Its first coordinate, for example, the radial displacement of the upper bearing is

(4. 6) $W_1 = X_1 + i X_2 = X\,e^{i\Phi_x},$

where $X(\Omega)$ is the numerical value of the displacement and $\Phi_x(\Omega)$ the phase angle between the radius vector of the exciting

unbalance and the resulting displacement vector, both of them
function of Ω. Fig. 8 shows the typical amplitude curves
$X(\Omega)$, $Y_0(\Omega)$, $Y_u(\Omega)$, $Z(\Omega)$ for the upper bearing, the upper
and the lower rotor end and the lower bearing.

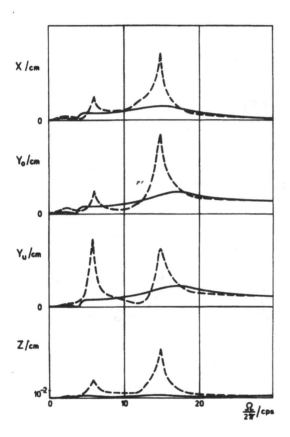

Fig. 8

There are four resonance peaks at about 2, 6, 12 and 15 hz
which are, because of damping, more or less distinct. The
critical speeds corresponding to these resonances are indicat-
ed in Fig. 7 by the intersection points of $\omega_k = \Omega$ with the frequency curves of the forward whirl. The reason why there are

only four critical speeds as compared to eight natural frequencies is detailed in Chapt. 3.3. Fig. 9 shows the polar plots
that correspond to the amplitude curves giving additional information on phase angles.

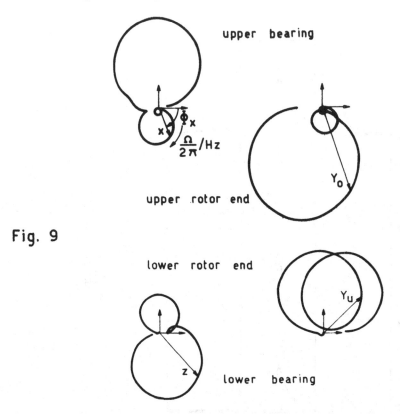

Fig. 9

4.2 Optimization of parameters.

The free parameters of the system have to be determined so that its dynamic behavior will be optimal in a sense
that has to be specified [13]. A measure of the quality of an
optimum is given by an optimization criterion that weights
the technical performance requirements in a suitable way. The

system has eight free parameters : the two damping coefficients of the bearings, the masses of the two bearings and the four elasticity coefficients of the support. The free parameters can be adjusted within a certain domain. The optimization problem consists principally of three parts :

1. Choice of a suitable optimization criterion;
2. Calculation of the criterion;
3. Determination of the optimal parameters. Independently of 1. and 2. the free parameters have to be determined in such a way that the criterion comes to an extremum.

This is done by a search procedure that samples the eight-dimensional space of parameters step by step and that approaches gradually better values of the criterion up to the immediate neighborhood of the optimum [14]. In the following the choice of a suitable criterion is dealt with to some detail, whereas the calculation of the criterion is restricted to the giving of literature and results.

Maximal degree of stability :

The degree of stability is the distance h of the imaginary axis from the next root of the characteristic equation for a stable system. The degree of stability h supplies a measure for the rate of decay of a transition motion. A large degree of stability results in a fast decaying of disturbances of the equilibrium position. The degree of stability

(4.7) $$h = - \text{Max}_i \{ \text{Re} \, \lambda_i \}$$

can be calculated from the eigenvalues of the real or complex

system (4.5). After that the maximal degree of stability in the

parameter space has to be determined. The set of parameters

for which the degree of stability becomes maximal is shown in

Fig. 10 and compared to other parameter sets.

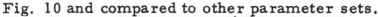

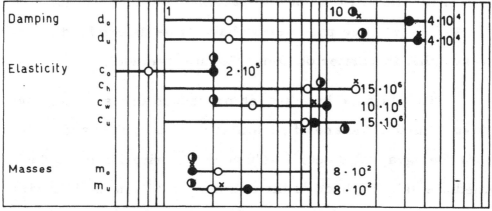

Fig. 10

—   domain of the free parameters
o   parameters for the real model
x   parameters for optical degree of stability
◑   parameters for minimal resonance amplitude (M R A)
●   parameters for minimal IE^2

The damping and elasticity coefficients lie in the upper part or

even at the end of the admissible domain. At first the degree of

stability for the model that had been used for experiments ini-

tially was 0.86. This value could be improved by optimization

up to 19.7. It had been calculated for a relatively low rotor

speed of $\Omega = 50 \, \text{hz}$. With rising rotor speed all the optimal

parameters tend to the end of the domain. The maximal degree
of stability is given by the damping of the least damped free
motion and for high rotor speed this is nutation or precession.
From Chapt. 3. 3. it is already known that the damping of nutation and precession decreases with rising rotor speed and this
principal tendency cannot be removed even by optimization, it
can only be mitigated.

The degree of stability characterizes decaying of free
motions at a certain rotor speed but it says nothing about their
amplitudes nor about the amplitudes of forced motions when
running, for example, through critical speeds. But from a
technical point of view a limitation of amplitudes is important.
Therefore two criteria are presented that entail a reduction of
the rotor displacements caused by disturbances.

Minimal resonance amplitude (MRA) :

The behavior of the rotor passing critical speeds in a
quasistationary way and its behavior against harmonic disturbances at any constant rotor speed can be described by the amplitude and phase curves (4. 6). Of special interest are the displacements y_0 and y_u of the upper and lower rotor end. It is
efficient to tune the system parameters in such a way as to
minimize the maximal amplitudes, i. e. in a certain domain
$\bar{\Omega}$ of the rotor speed the resonance curves have to show resonance peaks being as small as possible:

(4.8) $K = \text{Max} \left\{ \underset{\Omega \epsilon \bar{\Omega}}{\text{Max}} \; Y_u(\Omega), \; \underset{\Omega \epsilon \bar{\Omega}}{\text{Max}} \; Y_0(\Omega) \right\} \longrightarrow \text{Min}.$

The maximum of the resonance peaks is easily calculated from the amplitude curves (3.15) as they are merely functions of the rotor speed Ω and do not depend on time. As an
example the displacement of the upper rotor end is given by

$$y_{10} = y_1 + c\beta \qquad\qquad y_{20} = y_2 - c\alpha$$

or in complex form

$$\underset{\sim}{y}_0 = y_{10} + i\, y_{20}$$

and by (4.2) follows

$$\underset{\sim}{y}_0 = w_2 + c\, w_4 = \underset{\sim}{y} + c \odot.$$

From the particular solution (3.15) results

$$\underset{\sim}{y}_0 = (W_2 + c W_4)\, e^{i\Omega t}.$$

Finally the complex amplitude $W_2 + c W_4$ of the upper
rotor end can be split into the real amplitude $Y_0(\Omega)$ and the
phase angle $\Phi_{y0}(\Omega)$ so that

(4.9) $\underset{\sim}{y}_0 = Y_0\, e^{i(\Omega t - \Phi_{y0})}$

holds. Only the amplitude $Y_0(\Omega)$ is of interest in this connection. Here the domain for Ω includes all critical speeds.

The set of parameters which permits running up with minimal resonance peak is shown in Fig. 10 and is compared to other optimal sets. It is noteworthy that the damping does not go to the upper end of the domain and that the elasticity coefficient of the rotor shaft takes the smallest possible value. All the other parameters take similar values as given by the other criteria.

Fig. 8 shows amplitude curves with the lowest possible resonance peak together with the amplitude curves for the initial parameter set of the model. In this case the maximal resonance peak of the upper rotor end reduces to 29% of the initial maximum.

IE^2 - criterion.

For the continuous operation of the gyro at constant rotor speeed its behavior as caused by impulse-shaped disturbances may stand for its general behavior. The impulse response is very suitably characterized by the generalized integral of square error (IE^2-criterion). For an asymptotically stable multi-variable system it is defined by

$$IE^2 = \int_0^\infty u(t)\, B\, u(t)^\tau\, dt, \qquad (4.10)$$

where the state space vector u is given by

(4.11)
$$u = \begin{bmatrix} v \\ \dot{v} \end{bmatrix}$$

with v following from (4.1). By suitably choosing the weighting matrix B and the initial state space vector $u(o)$ a great variety of special requirements as to the character of the motion can be considered. Here it seems to be efficient to keep the rotor displacements small when an impulse acts on the rotor. Thus, the initial state space vector is chosen as

$$u(o) = \begin{bmatrix} 0,0,0,0,0,0,0,0 \; ; 0,0,0,1,0,0,0,0 \end{bmatrix}^T$$
(4.12)

and the weighting matrix has the form

$$B = \text{diag} \begin{bmatrix} 0,1,0,1,0,1,0,1 \; ; 0,0,0,0,0,0,0,0 \end{bmatrix},$$

(4.13)

where only the displacements of the rotor center of mass and the inclinations of the rotor are weighted.

The IE^2 criterion in the form (4.10) can not be computed numerically for the underlying system. In [15] it is shown that it can be transformed into

(4.14)
$$IE^2 = \frac{\det H^+}{2\, a_0 \det H}$$

Here the matrix H is the Hurwitzian of the state space equation

$$\dot{u} = A u \, ,$$

where A results from (4.1) and (4.11) and is given by

$$A = \left[\begin{array}{c|c} 0 & I_8 \\ \hline M^{-1}(D+G) & M^{-1}F \end{array} \right]$$

and a_0 is the coefficient of the highest power of the corresponding characteristic polynomial. The matrix H^+ differs from H only in the first column where the initial condition $u(0)$ and the weighting matrix B are taken into account.

An effect, not considered in the aforesaid criteria, is that of parameter sensitivity. For this reason some theoretical results could not be verified by experiment. As an example Fig. 11 (see next page) shows the IE^2-value depending on the damping parameter d_0. It is reasonable that the absolute optimum 0_1 for low damping is of practically no use and that the optimal damping parameter d_0 is determined by the relative but nevertheless true optimum 0_2. The corresponding IE^2-optimal parameters are given in Fig. 10. The improvement of the system behavior can be seen by comparing the impulse responses of the optimal system and of the initial model. Fig. 11 represents for both cases the motions of the bearings (x_1, z_1), the centre of mass of the rotor (y_1) and the upper ro-

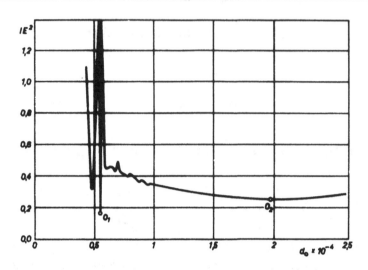

Fig. 11

tor end $\left(y_{10}\right)$.

The IE^2 – optimal set of parameters warrants a very
good behavior of the system in the case of continuous operation.
It has to be checked, however, how the system tuned in this way
behaves when the rotor runs up. Though the passing through
critical speeds is not weighted by the IE^2-criterion the maximal resonance peak is reduced and that to 45% of the
initial maximum. As compared to the running up with minimal
resonance amplitude as shown before, now a change for the
worse by a factor 1.7 has to be put up with in order to obtain
an optimal behavior at continuous operation.

4.3. Experiments.

The measuring technique that is used for studying and

controlling vibrations of high-speed rotors is part of the general vibration technique. One point of view which helps to distin-
guish devices and methods may be the following one : Depending
on the kind of application it can be distinguished between measuring technique for laboratory and for operation. In the first
case qualitatively new results are aimed at. Therefore multi-
purpose devices and methods on account of the expected variety
of effects have to be applied. In the latter case quantitative results on usually known properties of the system are often need-
ed in a very reliable, fast and exact way and all that during long
intervals. An example of it is the vibration control system for
the turbines of a power plant or the field of balancing machines.
Another point that may help to classify this measuring technique is the character of the vibrations, i.e., if they are mainly
deterministic or stochastic. In the first case the sources
of disturbance can often be exactly localized on account of their
discrete frequencies, for example rotor unbalances. Moreover,
discrete values like critical speeds or maximal amplitudes are
investigated. In the latter case the vibrations are noisy and interest is laid upon mean squares and spectra. These investigations go far into the area of acoustics. Here in each case the
first criteria are considered : laboratory measuring of deterministic rotor vibrations. Referring to the known example of
the elastically supported rotor (Chapt. 4.1) the systematic
manner of the experimental set-up is shown, and some special

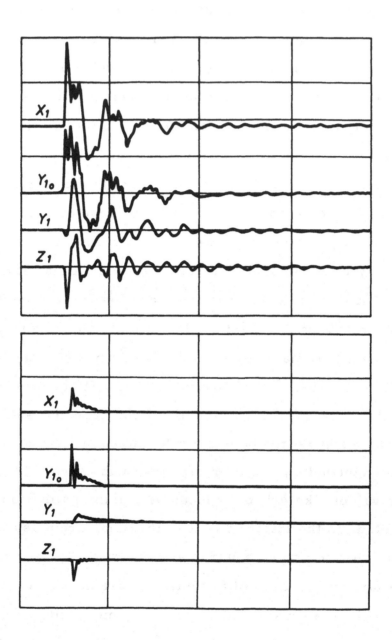

Fig. 12

aspects are emphasized which are, in the opinion of the author, of general interest.

In general, the aim and purpose of an experiment is to check and to verify a theoretical model being the basis for a theoretical investigation of a system by means of the real model, or relations and data have to be measured which now on their parts permit to establish a theoretical model. To this end, experiments on a gyroscopic model corresponding to the theoretical model of Fig. 6 are effected and evaluated. The connection of the different problems arising herein is shown in the block-diagram of Fig. 13.

The measured variables of the gyroscopic model are of mechanical nature being displacements and angular velocities. They have to be transformed by suitable pickoffs into electrically measured values. In order to obtain a defined connection between the original mechanical and now electrical values the measuring program has to care for calibration procedures. Hence, for the purpose of calibration a defined input signal acts on the model. For analyzing the measured values as well and for the parameter identification a defined input signal, for example a harmonic vibration, is very useful. Parameter identification stands here for the determination of those parameters which describe the theoretical model (4.3), for example moments of inertia, damping and elasticity.

The measured values coming from the pickoffs are stor-

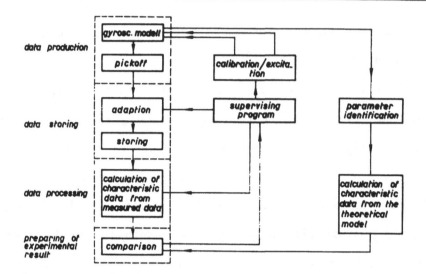

data production

data storing

data processing

preparing of experimental result

Fig. 13

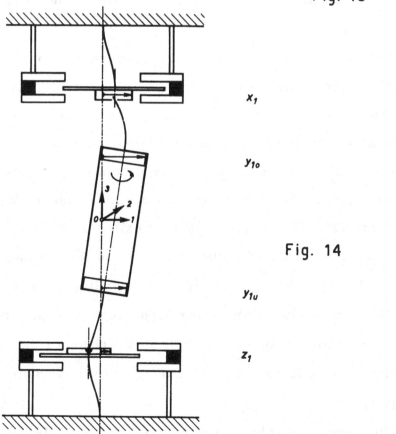

x_1

y_{1o}

Fig. 14

y_{1u}

z_1

ed depending on the supervising program and if necessary before that the signal is transformed and electrically adapted to the storage unit. Data processing means that characteristic properties of the system like natural frequencies or amplitude curves are determined from the measured data. A comparison between these experimentally achieved results and the results calculated from the theoretical model in the end leads to a final result that will obviously influence the way of continuing the whole program. In the following, the different blocks of fig. 13 are briefly dealt with.

Data production: Fig. 14 shows the principal construction of the mechanical model. The horizontal displacements of the upper and the lower bearing, the upper and the lower rotor are measured, and that at the circumference of the rotor and the bearings in the 1- and 2- direction, respectively.

By these components the total displacement is fully described. Altogether there can be measured 8 signals which permit the representation of the system with 8 degrees of freedom. However, it shall be mentioned as it turns out to be that for answering the most questions it suffices to measure only the displacements in the 1 or 2-direction as the motions in these directions are quite similar. Then, there are only four signals to be measured.

The displacements are measured by opto-electrical pisckoffs. The rotor acts as a slit diaphram between a source of light and a detector and the quantity of light falling into the

detector is a measure for the displacement. The advantage of
this method of measuring deviations of a rotating body free of
contact, with small time-constants and low noise is obvious.

The angular rotor velocity is measured opto-electrically, too. The circumference of the rotor is divided into black
and white sections which are sampled when the rotor turns
and thus the pickoff yields impulses from which the rotor veloc
ity and the angle of position of the rotor are determined.

Calibration : The calibration to which great attention should be
paid is effected statically by special calibration screws, so
that any displacement corresponds to a measured value.

Excitation : A systematical investigation of the system behavior
not only requires the knowledge of the output signals of the gyroscopic model but as well the knowledge of the input signals.
A symple possibility of producing an input signal is the harmon
ic excitation of the model by the rotor unbalance itself. But it
can only cause resonance if the rotor velocity coincides with a
natural frequency of the system, in general, only when the rotor
runs up or down. Furthermore, these natural frequencies differ from those which are valid when the rotor is at full operation
al speed. A harmonic excitation which does not depend on the
rotor velocity is obtained by means of a special excitation device. It consists of a small direct current motor driving an unbalance and is connected to the upper bearing. The vector of the
disturbance forces in (4.3) for the ideally balanced rotor now

changes into

$$\bar{r}\,e^{i\Omega t} = \left[u\,\Omega^{*2},\,0,\,0,\,0\right]^{T} e^{i\Omega^{*}t}, \qquad (4.15)$$

where u is the unbalance and Ω^{*} the angular velocity of this
unbalance. The angular velocity and the position of the unbalance is measured opto-electrically.

   As an example of non-harmonic excitation the step or the
impulse acting on the rotor is worth considering. From the step
transition functions the damping of the bearings can be determined. The impulse response permits checking directly the ef-
ficiency of an optimal parameter adjusting (Chap. 4.2). In order to
excite the rotor by an impulse in a reproduceable way small plas-
tic balls have been shot from a toy pistol on to the turning rotor.

   Theoretically these impulse responses contain all
facts about the behavior of the system at all disturbance frequencies. However, the expenditure for the evaluation of the
impulse response is considerable, and, moreover, for a somewhat higher damping of the system the impulse response is any-
thing but distinct so that the errors of data processing will become too large. A harmonic excitation has the advantage as
compared to an impulse that in a controllable way more energy
can be pumped into the system and therefore more distinct motions of the systems can be forced. An investigation of the rotor
by means of stochastic input signals had been omitted as it requires expensive devices for a defined excitation as well as for
evaluation of data. Moreover, extremely long measuring times

have to be taken into account in order to get somewhat exact infor

mation about low-frequency components of the stochastic vibration.

Transformation and adaption : A comparison between the results

of the theoretical and the experimental investigations is only

possible if the relations between the measured variables and the

variables which describe the theoretical model are known. If

these relations are simple the necessary transformations can

already be done with the measured data. This can remarkably

reduce the amount of data or save later transformations of the

final results.

The variables that describe the displacements of the

bearings are the same for theory and experiment. But as to the

rotor its position is theoretically given by the displacements of

the center of mass and the inclination of the rotor and it is ex-

perimentally given by the displacements of the geometrical axis

of symmetry of the rotor and that at the upper and at the lower

rotor end. The relation between the measured variables

$$n = \left[n_{x1} , \; n_{01} , \; n_{u1} , \; n_{z1} , \; n_{x2} , \; n_{02} , \; n_{u2} , \; n_{z2} \right]^T ,$$

i. e. the measured displacements of the upper bearing, the upper

and the lower rotor end and the lower bearing in the direction of

the 1- and 2- axis, respectively, and the vector of the system

variables is given by

(4.16) $n = Pv + q ,$

where

$$P = \begin{bmatrix} \begin{array}{cccc} 1 & 0 & 0 & 0 \\ 0 & 1 & 0 & c \\ 0 & 1 & 0 & c_1 \\ 0 & 0 & 1 & 0 \end{array} & \Large 0 \\ \hline \Large 0 & \begin{array}{cccc} 1 & 0 & 0 & 0 \\ 0 & 1 & 0 & c \\ 0 & 1 & 0 & c_1 \\ 0 & 0 & 1 & 0 \end{array} \end{bmatrix}$$

$$q = \begin{bmatrix} 0, & -a\cos\Omega t, & 0, & 0 & ; & 0, & -a\sin\Omega t, & 0, & 0 \end{bmatrix}$$

and c, c_1 are the distances from the centre of mass to the upper and lower rotor ends. By means of an analogue computer using only amplifiers and potentiometers (4.16) can easily be effected. The measured values for the rotor velocity have to be transformed as well. They are given as a sequence of impulses produced by the photo-electric pickoff. By means of a hybrid analogue computer this signal is transformed into a direct current proportional to the angular velocity and furthermore a correlated sin- and cos-wave of constant amplitude is shaped. This will be used as a reference signal for the following data processing. It shall be mentioned that the hybrid analogue computer has successfully been applied as a very suitable and versatile measuring device. With its help it has been possible and very efficient to omit all data storing and to evaluate the data at once.

Data processing: Representation of the time-dependance of the gyroscopic motion is only suitable for preliminary experiments. Fig. 15 shows the vibrations of the system in the 1-direction at the nutation frequency of 15 hz.

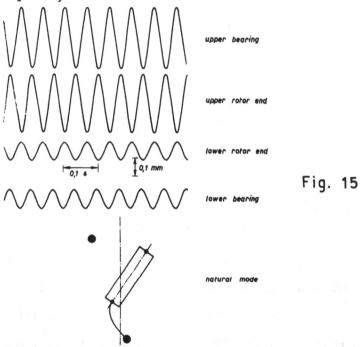

Fig. 15

The phase-relations of the vibrations are used to construct the shown natural mode of nutation. The information contained in the amplitude-time behavior with respect to amplitude frequency characteristics, dependance on acceleration of the rotor, and phases are often difficult to recognize and have to be evaluated by a suitable data processing. The amplitude curves and polar plots are calculated by a hybrid analogue computer directly from the measured data and simultaneously plotted. As an example Fig. 16 shows a calculated and measured amplitude frequency response of the upper bearing.

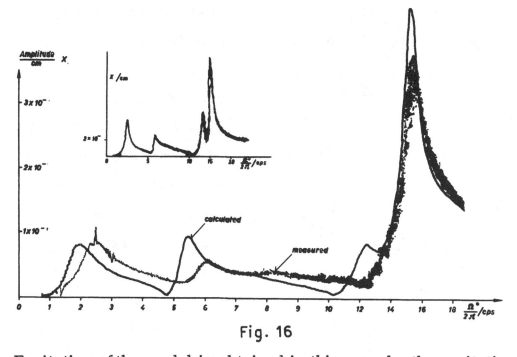

Fig. 16

Excitation of the model is obtained in this case by the excitation device corresponding to (4.15). The rotor itself does not turn. The resonance peak at this critical speed of about 12 cps is only weakly shaped. It can not be seen in the measured curve but it comes out quite distinctly if the vibrations are less damped as it is shown by the scaled down resonance curve. Figures 17 and 18 show polar plots corresponding to Fig. 16. The resonances are more easily recognized on account of their loops and edges. The measured polar plot had been recorded twice in order to check the repeatability of the results. The deviations from the normally smooth curve at about 14 hz result from nonlinear transitions that occur at large amplitudes. Resonance frequencies,

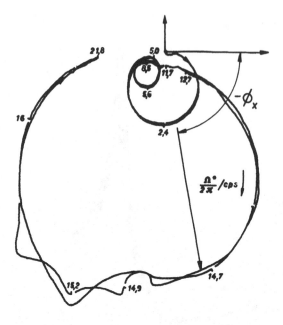

Fig. 17

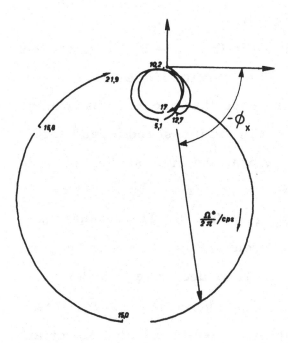

Fig. 18

phases and amplitudes of the vibrations did correspond fairly well in calculation and experiment. Thus, the assumptions which had been stated for the theoretical model have been justified. Deviations from these assumption and thus from the calculated behavior can occur, for example, by the following reasons: if the coupling between the transversal motions gets too strong, due to large deflections of the rotor, and if the restoring characteristics of the elastis support are no more linear, the nonlinear effects have to be considered (see Chap. 6.1). The stability of the rotation can be effected if the internal damping, for example due to elastic deformations of the rotor itself, has to be taken into account (see Chapt. 6.2).

5. Balancing.

Definition : Balancing is a process where the moments and forces acting on a rotating body and caused by unbalances are used to measure the unbalance and to remove it. By static balancing a static unbalance is removed, i.e. the centre of mass of the body is laid on a given axis of rotation. By dynamic balancing the dynamic unbalance is removed. A given axis of rotation becomes a main axis of inertia of the body. For defining the term unbalance let a disc be equipped with an additional mass, Δm, at its circumference (Fig. 19 a) (see aside):

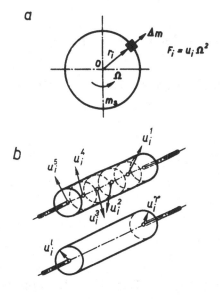

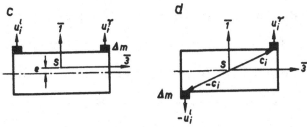

Fig. 19

The product

(5.1) $u_i = \Delta m \cdot r_i$

is the unbalance. It is a vector and has the same direction as
the centrifugal force

(5.2) $F_i = \Omega^2 \cdot u_i$

which acts on the additional mass of the rotating disc. The additional mass leads to a displacement e_i of the center of mass
and thus

(5.3) $u_i = \Delta m \cdot r_i = (m_s + \Delta m)e_i = m \cdot e_i$

holds, where m is the total mass. Hence, the centrifugal force
is

$$F_i = \Omega^2 m\, e_i$$

and its numerical value is

(5.4) $F = \Omega^2 m\, e .$

Now let the cylindrical rotor of Fig. 19b be composed
of several unbalanced discs. The different unbalances result
according to the rules of statics for example in two unbalances
u_i^r and u_i^ℓ in two chosen planes of the body, here in the
right and the left planes of the rotor ends. By these two unbal
ances the state of unbalance of the rotor is definitely given.

The unbalances can be removed in each of these planes (balancing planes) independently of one another; for example by removing the additional mass or by adding an equal mass to the opposite side. The two unbalances having different direction and largeness can always be split up into a static and a dynamic component. In the case of a static unbalance (Fig. 19c) the components of the centrifugal forces acting on the unbalances u_i^ℓ and u_i^r are such that they result into one centrifugal force F_i passing through the center of mass. For Fig. 19c this means that with $u_i^\ell = u_i^r [u, 0, 0]^T$ the coordinate F_1 of the centrifugal force (in the body fixed 1 2 3 coordinate-system) is

$$F_1 = 2 \Omega^2 u = \Omega^2 m e. \qquad (5.5)$$

In the case of a dynamic unbalance (Fig. 19d) components of the centrifugal forces acting on the unbalances are such that they result in a free moment. For Fig. 19d this means that with $u_i^r = - u_i^\ell = [u, 0, 0]^T$ and $c_i = [r, 0, c]^T$ the resulting coordinate M_2 of the free moment is

$$M_2 = 2 c \Omega^2 u = 2 c r \Delta m \cdot \Omega^2 = E \Omega^2. \qquad (5.6)$$

The product of inertia E follows from the definition

$$E = \int x_1 x_3 \, dm = 2 c r \Delta m, \qquad (5.7)$$

where x_i denotes the vector to a point of mass of the rotor.
The small angle α between the main axis of inertia and the
geometrical axis of symmetry (3-axis) is then given by

(5.8)
$$\alpha \simeq \frac{E}{C-A} = \frac{2cu}{C-A} \, ,$$

where C and A are inertia moments with respect to the 3- and
1- axis.

When balancing the rotor its bearings are elastically
supported so that the forces and moments caused by unbalances
lead to vibrations of the bearings. As the stiffness of the bearings is chosen in such a way that the natural frequencies of the
system are sufficiently below the operational velocity the rotor
tends to rotate about a principal axis of inertia going through
the centre of mass. Therefore in the case of a static unbalance
the vibrations of the bearings are equally phased, i.e. they are
displaced in the same direction whereas in the case of dynamic
unbalance the vibrations of the bearings are out of phase. The
vibrations of the bearings are measured by externally fixed
pickoffs. Let them measure for example the signals

(5.9)
$$n^r = a^r \cos(\Omega t - \Phi^r) + s(t)$$
$$n^l = a^l \cos(\Omega t - \Phi^l) + s(t)$$

on the right and the left bearing, respectively, where a^r and
a^l denote the amplitudes of the vibrations which occur at the
rotor velocity Ω. The phase angles Φ^r and Φ^l denote the angles

between the unbalances u_i^r and u_i^ℓ and a body-fixed reference direction, for example the 1-axis. The angular position of this reference is measured as well for example by photo-electrically sampling a mark on the circumference of the rotor. This signal serves to generate a $\sin \Omega t$ and a $\cos \Omega t$ that later on will be correlated with the output signals (5.9). The additional component $s(t)$ in (5.9) is composed of disturbing vibrations with frequencies unlike Ω and it can be expressed as :

$$s(t) = \sum_n a_n \cos(\Omega_n t - \Phi_n). \qquad (5.10)$$

The correlation procedure that is applied in order to evaluate the unbalances is known - depending on the specific domain where it is used - as evaluation of polar plots, as Fourier-analysis or watt-metric procedure. The harmonic of the vibration of the bearing that results from unbalances is according to (5.9) for the right bearing (the further evaluation is restricted to the right unbalance)

$$a^r \cos \Phi^r \cos \Omega t + a^r \sin \Phi^r \sin \Omega t,$$

where

$$a^r \cos \Phi^r = \frac{\Omega}{\pi} \int_0^{2\pi/\Omega} n^r \cdot \cos \Omega t \, dt$$

$$\qquad (5.11)$$

$$a^r \sin \Phi^r = \frac{\Omega}{\pi} \int_0^{2\pi/\Omega} n^r \cdot \sin \Omega t \, dt$$

are calculated from the measured signal n^r by correlating it
to the harmonic reference signal. By way of integration the
disturbances are averaged out. The values determined
by (5.11) are the coordinates of the displacement a_i^r of the
right bearing in rotor-fixed coordinates, and a^r is proportion
al to the numerical value of the right unbalance u_i^r, and Φ^r char
acterizes the angular position of the unbalance. The proportion
ality factor can be determined in the following way : The meas·
ured displacement a_i^r is split up into two components a_i^{rs} and
a_i^{rd} caused by static and dynamic unbalance, respectively,
using to this aim the displacement a_i^ℓ of the left rotor side as
well

$$a_i^r = \frac{1}{2}\left[a_i^r + a_i^\ell\right] + \frac{1}{2}\left[a_i^r - a_i^\ell\right],$$

(5.12)

$$a_i^r = a_i^{rs} + a_i^{rd}.$$

Now for static unbalance the numerical value of the
displacement is equal to the eccentricity

$$a^{rs} = e$$

and therefore with (5.5) the static unbalance is

(5.13) $$u^{rs} = \frac{m}{2}\, a^{rs}.$$

For the dynamic unbalance the numerical value of the
displacement is

(5.14) $$a^{rd} = c\alpha,$$

where c is the distance between the rotor end and the center
of mass of the rotor and α is the inclination of the principal
axis of inertia with respect to the axis of symmetry of the rotor
as it was calculated for a special position of the coordinate system in (5.8). Anyway

$$\alpha = \frac{2 c u^{rd}}{C-A}, \tag{5.15}$$

u^{rd} being the dynamic unbalance and hence from (5.14)

$$u^{rd} = \frac{C-A}{2 c^2} a^{rd}. \tag{5.16}$$

Finally, the composite unbalance on the right rotor end
is

$$u_i^r = u_i^{rs} + u_i^{rd} \tag{5.17}$$

or, with the proportionality factors of (5.13) and (5.16)

$$u_i^r = \frac{m}{2} a_i^{rs} + \frac{C-A}{2 c^2} a_i^{rd}, \tag{5.18}$$

where on the right side of the equation are given constants and
values which have been determined from measured vibrations.

Of course, there will be still other parameters which
influence the indication of a balancing machine, even when balancing rigid rotors. The calibration of these machines, the influence of unequal stiffness and inertia properties, the error

analysis, the technical requirements on balancing quality, the balancing of elastical rotors and the balancing at resonance are problems that go far beyond the aim of this lecture and are dealt with in special literature [17] .

6. Influence of different parameters.

6.1. Running through critical speeds.

A gyroscopic system, for example an elastically supported gyro, passes critical speeds when the rotor runs up or
down. The vibrations of the rotor and the bearings become
large and are caused by the unavoidable unbalances of the rotor.
Investigations of one-degree-of freedom vibrators have shown
[27] that the resonance amplitude can be reduced by rapidly
running through resonance. Now it is of interest if and to what
extent this result is valid for systems with several degrees of
freedom. As an actual example the elastically supported gyro
(Chapt. 4) may serve for this investigation but most of the results can be generalized.

The equation of motion is not derived in detail here
but it is quite plausible that it given by an equation equal to
(4.1) with the exception of the gyroscopic matrix G which is
no longer constant and the excitation $f(t)$ which is no longer
harmonic, and that on account of the variable rotor velocity
$\Omega = \dot{\gamma} \neq$ const. The corresponding state space equation has the
form

$$\dot{u} = \left[A(\dot{\gamma})\right] u + h(\dot{\gamma}), \qquad (6.1)$$

where the state space vector

$$u = \begin{bmatrix} v \\ \dot{v} \end{bmatrix},$$

the coefficient matrix

$$A(\dot{\gamma}) = \left[\begin{array}{c|c} 0 & I_8 \\ \hline M^{-1}[D+G(\dot{\gamma})] & M^{-1}F \end{array}\right],$$

the excitation

$$h(\dot{\gamma}) = \left[\begin{array}{c} 0 \\ f(\dot{\gamma}) \end{array}\right].$$

In addition to (6.1) the equation for the rotation $\dot{\gamma}$ about the rotor axis holds and has the linearized form

(6.2) $$c\ddot{\gamma} = T,$$

where T is the driving torque. But the preceeding linearization is permitted only if the coupling between the transverse motions and the rotation $\dot{\gamma}$ is small. This is the case which occurs when the vibrations are sufficiently small and the driving torque sufficiently large. For the sake of simplicity, let the torque be constant and then the rotor velocity

(6.3) $$\dot{\gamma} = \frac{T}{C}t$$

is linearly time-dependent. Then (6.1) has time-variable coefficients and can not be analytically solved. A stability study of (6.1) can of course be performed which holds as well for general gyroscopic systems (3.14) and it turns out that the nonstationary passing of critical speeds remains bounded and therefore

stable if the vertical position of the gyro is asymptotically stable at any constant rotor-velocity up to the operational speed,
if the disturbance $h(\dot{\gamma})$ is bounded and if the passing of the critical speeds is finished within a finite time interval. But this
statement does not say much about amplitudes that arise when
the rotor runs more or less rapidly through the critical speeds.
This can be investigated only by numerical solutions for given
parameters and this was done by the aid of a hybrid analogue
computer. Some specific properties of these solutions may be
of general interest. Fig. 20 shows the amplitude curves of the

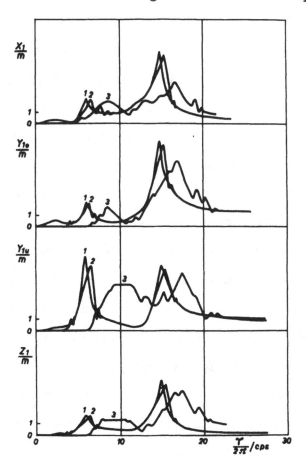

Fig. 20

vibrations in the direction of the 1-axis and it clearly shows
the influence of the rotor acceleration on the position and height
of the resonance peaks. The larger the acceleration is, the
more the peaks deviate from their position at stationary operation. When running up the peaks shift to higher frequency values, when running down to lower frequencies (Fig. 21). The nu

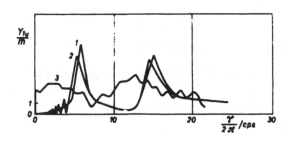

Fig. 21

merical values of the ac-
celerations are: 1: 0,1
cps/s; 2: 1 cps/s; 3: 10
cps/s for the running up
as well as for the running
down. The amplitudes them
are referred to a reference
amplitude m and give rise
to the supposition that

they reduce when the rotor acceleration increases and that the
resonance amplitudes are less dominant the shorter the time
is for passing the resonances. But the amplitudes can not be
reduced at will as the reduction is not at all proportional to
acceleration. Furthermore, if the damping of the system is
relatively high,as   is typical with optimally tuned parameters,
acceleration hardly influenced the height of resonance peaks
as Fig. 22 shows. Therefore the fast running through critical
speeds is not, in any case, the best way to reduce resonances.
It seems to be better to consider stationary amplitude-frequen-

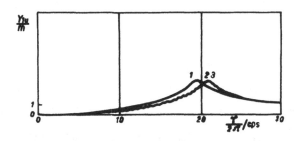

Fig. 22

cy-responses and to reduce them by optimizing the free parameters (Chapt. 5.2).

The quasistationary running up is taken as typical but
in reality the rotor cannot run up slowly at will, at least not
if it is driven by a small constant torque. Then, the coupling
between the transverse motion and the rotational velocity
can be so strong that the running up stops at a certain velocity.
It would be very difficult to calculate the exact time-dependance of the rotor velocity from the complete nonlinear equations. However, the minimal driving torque which is necessary
in order to avoid this "catching" of the running-up can be determined from energy relations. The total power which is fed
to the system by the rotor drive is

$$P = T \dot{\gamma} \tag{6.4}$$

and serves as power P_a for acceleration of the rotor, as vibrational power P_v for building up the vibrations of the sys-

tem and it is partly dissipated by the bearings as damping power P_d :

(6.5) $$P = P_a + P_v + P_d .$$

Now the running-up is caught at a certain velocity (then $P_a = 0$ holds), if the stationary vibration at this velocity is built up (then, $P_v = 0$ holds) and if the total power fed to the system is dissipated. Hence,

(6.6) $$P = P_d$$

remains.

The damping power can be easily calculated from (4.1)

(6.7) $$P_d = \dot{v}^T D \dot{v}$$

and is shown (for the numerical values of the real model) together with the total power (6.4) in Fig. 23.

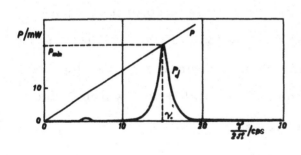

Fig. 23

The intersection point of the two curves leads to the rotor velocity $\dot{\gamma}_0$ The minimal driving torque necessary in order to avoid "catching" follows from Fig. 23 and (6.4)

(6.8) $$T_{min} > P_{min} / C \dot{\gamma}_0 .$$

This also implies that the rotor cannot pass resonances uniformly slower than with the acceleration

$$\ddot{\gamma}_{min} > P_{min} / C \dot{\gamma}_0 \qquad (6.9)$$

according to (6.2) and (6.8). The experiment did verify this result.

The slowly running-up of the rotor had revealed another nonlinear effect appearing at experiments with low damping and large unbalances. The running-up is not only caught at a certain resonance frequency but it is caught in a stable limit-cycle as is shown in Fig. 24.

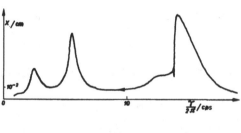

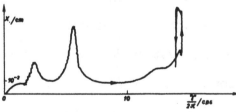

Fig. 24

The amplitude of the vibration of the upper bearing jumps at two distinct rotor velocities from a low to a high level and vice versa. When the rotor runs down as a result of friction only, the periodic limit-cycle does not occur. It turns out that the restoring force between upper bearing and upper rotor end is not proportional to deflection but is underlinear. This single nonlinearity leads not only to

a nonlinear behavior of the afflicted coordinates of the upper
bearing and rotor end but causes similar effects in all other
coordinates as well, a phenomenon that has been reported by
TONDL [16] as well. But as to the pumping effect of the limit
cycle nothing seems to be known yet about its mechanism.

6.2. Internal damping.

Internal damping means that the damping forces are
caused by the deformation of the vibrational system itself. It
is a so-called structural damping, i. e., by internal friction of
the material of the shaft when it is bending. The effect has been
studied by several authors [1, 3, 18, 19, 20] . The main problem
in studying the dynamic behavior under the influence of internal
damping does not lie in calculating it but in establishing the cor-
rect theoretical model for describing the kind of damping. There
are nonlinear damping characteristics causing self-excited vibrations, there is hysteretic damping and damping which depends on the model for the material deformation differing from
the ideal Hook behavior. As long as the system does not contain
rotating parts and is a pure vibrational system the effect of external and internal damping is quite similar : free motions are
damped, forced motions remain bounded. This behavior changes
significantly for the system with rotating parts: external damping acts as before and it supports stability at all rotational
speeds, but internal damping causes instability if the rotor vel-

ocity exceeds a certain value. This effect is not specific to
gyroscopic systems : it occurs as well when gyroscopic forces
are neglected. But - in order to show the relation between gyroscopic parameters and the destabilizing internal damping the
example of Chapt. 2, the rotor on overhung shaft, is used again.
The equations of motion (2. 3) must now be completed:

$$m\ddot{y}_i = G_i + P_i + P_i^I + P_i^E$$

$$I_{ij}\dot{\Omega}_j + \varepsilon_{ijk}\Omega_j I_{kl}\Omega_1 = M_i + M_i^I + M_i^E,$$

(6. 10)

where these additional terms P_i^I, M_i^I and P_i^E, M_i^E are

caused by forces and moments of internal and external damping,
respectively. It is assumed that the center of mass S (Fig. 1)
coincides with the axis of symmetry that

$$y_i = u_i$$

(6. 11)

holds,as for this stability study the disturbances caused by unbalance need not be considered. The elastic restraints corresponding to (2. 5) are

$$P_i = - \begin{bmatrix} by_1 - c\beta \\ by_2 + c\alpha \\ -mg \end{bmatrix} \qquad M_i = - \begin{bmatrix} a\alpha + cy_2 \\ a\beta - cy_1 \\ 0 \end{bmatrix}. \qquad (6. 12)$$

The external damping arises for example by moving
the rotor in a viscous medium and thus producing forces and

moments which are proportional to the velocities of displacements and inclinations with respect to the fixed 1 2 3 coordinate
system :

$$(6.13) \quad P_i^E = - \begin{bmatrix} d_{11}\dot{y}_1 - d_{12}\dot{\beta} \\ d_{11}\dot{y}_2 + d_{12}\alpha \\ 0 \end{bmatrix}, \quad M_i^E = - \begin{bmatrix} d_{22}\dot{\alpha} + d_{21}\dot{y}_2 \\ d_{22}\dot{\beta} - d_{21}\dot{y}_1 \\ 0 \end{bmatrix}.$$

The damping forces and moments depend quite similar
ly on the superimposed influences of displacements and inclina
tion as do the elastic restraints.

The internal damping arises from the motion of the
rotor with respect to the rotor-fixed $\bar{1}\,\bar{2}\,\bar{3}$ - coordinate system.
Let $y_i = [y_1, y_2, 0]^T$ be the vector of displacement and $n_i = [\beta, -\alpha, 0]^T$ the vector of inclination, then the deformation
velocities with respect to the rotor, expressed in the fixed 1
2 3 coordinates system are

$$(6.14) \quad \begin{aligned} \overset{*}{y}_i &= \dot{y}_i - \varepsilon_{ijk}\Omega_j y_k, \\ \overset{*}{n}_i &= \dot{n}_i - \varepsilon_{ijk}\Omega_j n_k, \end{aligned}$$

where the star denotes the time derivation with respect to the
rotor-fixed coordinate system which rotates with $\Omega_i = [0,0,\Omega]^T$.

With suitable proportionality constants

$$P_i^I = - \begin{bmatrix} k_{11}\overset{*}{y}_1 - k_{12}\overset{*}{\beta} \\ k_{11}\overset{*}{y}_2 + k_{12}\overset{*}{\alpha} \\ 0 \end{bmatrix}, \quad M_i^I = - \begin{bmatrix} k_{22}\overset{*}{\alpha} + k_{21}\overset{*}{y}_2 \\ k_{22}\overset{*}{\beta} - k_{21}\overset{*}{y}_1 \\ 0 \end{bmatrix} \quad (6.15)$$

holds, or with (6.14)

$$P_i^I = - \begin{bmatrix} k_{11}(\dot{y}_1 + \Omega y_2) - k_{12}(\dot{\beta} - \Omega\alpha) \\ k_{11}(\dot{y}_2 - \Omega y_1) + k_{12}(\dot{\alpha} - \Omega\beta) \\ 0 \end{bmatrix},$$

$$M_i^I = \begin{bmatrix} k_{22}(\dot{\alpha} + \Omega\beta) + k_{21}(\dot{y}_2 - \Omega y_1) \\ k_{22}(\dot{\beta} - \Omega\alpha) - k_{21}(\dot{y}_1 + \Omega y_2) \\ 0 \end{bmatrix}. \quad (6.16)$$

Now (6.12), (6.15), (6.16) are substituted into the e-quation of motion (6.10). Arranging of terms in such a way that the variables form the vector

$$v = \begin{bmatrix} y_1, \beta, y_2, -\alpha \end{bmatrix}^T \quad (6.17)$$

leads to the matrix form

$$M\ddot{v} + (D + G)\dot{v} + (F + E)v = 0, \quad (6.18)$$

where

$$M = \text{diag} \begin{bmatrix} m, A, m, A \end{bmatrix},$$

$$D = \begin{bmatrix} \begin{matrix} d_{11} + k_{11} & , -(d_{12} + k_{12}) \\ -(d_{12} + k_{12}), & d_{22} + k_{22} \end{matrix} & 0 \\ 0 & \begin{matrix} d_{11} + k_{11} & , -(d_{12} + k_{12}) \\ -(d_{12} + k_{12}), & d_{22} + k_{22} \end{matrix} \end{bmatrix},$$

$$G = \begin{bmatrix} 0 & \begin{matrix} 0 & 0 \\ 0 & c\,\Omega \end{matrix} \\ \begin{matrix} 0 & 0 \\ 0 & -c\,\Omega \end{matrix} & 0 \end{bmatrix},$$

$$F = \begin{bmatrix} \begin{matrix} b & -c \\ -c & a \end{matrix} & 0 \\ 0 & \begin{matrix} b & -c \\ -c & a \end{matrix} \end{bmatrix}, \quad E = \begin{bmatrix} 0 & \begin{matrix} k_{11}\Omega, & -k_{12}\Omega \\ -k_{12}\Omega, & k_{22}\Omega \end{matrix} \\ \begin{matrix} -k_{11}\Omega, & k_{12}\Omega \\ k_{12}\Omega, & -k_{22}\Omega \end{matrix} & 0 \end{bmatrix}.$$

The stability criterion of Chapt. 2.2, requiring positive definiteness of the restoring matrix, cannot be applied here immediately, as the restoring matrix $E + F$ is unsymmetrical now due to the nonconservative forces Fv. It can be shown, however, that under certain symmetry-conditions these forces are only apparently nonconservative and can be removed by a suitable coordinate transformation [21, 22]. In order to show this let $d_{12} = d_{21} = 0, d_{11} = d_{22} = d$ and $k_{12} = k_{21} = 0$, $k_{11} = k_{22} = k$. Now

$$v = Px, \tag{6.19}$$

where P is an orthonormal transformation matrix, is substituted into (6.18). The matrix P is determined from the condition that the skew-symmetrical part of the restoring matrix vanishes; i. e., from

$$\dot{P} + D^{-1} E P = 0, \tag{6.20}$$

which leads to

$$P = \begin{bmatrix} \cos vt \cdot I & -\sin vt \cdot I \\ \sin vt \cdot I & \cos vt \cdot I \end{bmatrix}$$

where $v = k\Omega/(k+d)$ and I is a 2 x 2 identity matrix. Using the relation $\dot{P}P^T = N$ and $\ddot{P}P^T = NN$ with

$$N = v \begin{bmatrix} 0 & -I \\ I & 0 \end{bmatrix}$$

eq. (6.18 reduces to

(6. 21) $M\ddot{x} + D\dot{x} + (G + 2MN)\dot{x} + (K + MNN + GN)x = 0$.

A necessary and sufficient stability condition is that the now symmetrical restoring matrix $(K + MNN + GN)$ be positive definite. This means that the eigenvalues of the matrix

$$\begin{bmatrix} 1 - \Omega^2/\Omega_1^2 & -c/b \\ -c/a & 1 - \Omega^2/\Omega_2^2 \end{bmatrix}$$

have to be positive, where $a > 0$, $b > 0$ and

$$\Omega_1^2 = \frac{b}{m}\left(\frac{d+k}{k}\right)^2, \quad \Omega_2^2 = \frac{a}{A - C(d+k)/k}\left(\frac{d+k}{k}\right)^2.$$

This leads to the stability conditions

(6. 22) $1 - \Omega^2/\Omega_1^2 > 0$,

(6. 23) $1 - \Omega^2/\Omega_2^2 > 0$,

(6. 24) $\left(1 - \Omega^2/\Omega_1^2\right)\left(1 - \Omega^2/\Omega_2^2\right) - c^2/ab > 0$.

As a consequence of these three conditions the motion of the rotor is then and only then stable if the rotor speed is $\Omega < \Omega_s$ with

$$\Omega_s^2 = \frac{\Omega_1^2 + \Omega_2^2}{2} - \sqrt{\left(\frac{\Omega_1^2 + \Omega_2^2}{2}\right)^2 - \left(1 - \frac{c^2}{ab}\right)\Omega_1^2 \Omega_2^2}\,. \qquad (6.25)$$

If there is no internal damping k the motion is stable for any rotor speed Ω as long as $ab - c^2 > 0$. It is interesting to see that (6.23), characterizing a necessary condition for stable nutation, is always fulfilled if

$$A > C(d+k)/k\,, \qquad (6.26)$$

i.e., if the rotor is sufficiently disk-like. Furthermore it is desirable to make the external damping   much larger than the internal damping k in order to extend the stable range of rotor speed.

Further insight into the influence of the gyroscopic forces on the stability is given by looking at the damping-behavior of the highest eigenvalue of the system (6.18) that characterizes the damping of the nutation. Using the procedure outlined in Chapt. 3 the order of the system is reduced by using complex variables. Then the characteristic equation for the eigenvalues $\lambda_k = \delta_k \pm i\omega_k$ is

$$mA\,\lambda^4 + \lambda^3\left[-imC\Omega + m(d_{22} + k_{22}) + A(d_{11} + k_{11})\right] +$$
$$+ \lambda^2\left[\Omega(-iCd_{11} - iCk_{11} - imk_{22} - iAk_{11}) + \ldots\right] + \ldots = 0\,. \qquad (6.27)$$

For high rotor speed the comparison of (3.7) and (6.27) leads to the following relations: the parameter H denoting the rotor moment of momentum is replaced by Ω denoting the rotor speed and

$$b_m = Am$$
$$g_{m-1} = -mC$$

(6.28)
$$a_{m-1} = m(d_{22} + k_{22}) + A(d_{11} + k_{11})$$
$$h_{m-2} = 0$$
$$g_{m-2} = -C(d_{11} + k_{11}) - mk_{22} - Ak_{11}.$$

Then from (3.9) the nutation frequency follows

(6.29)
$$\omega_N = \mu\Omega = g_{m-1} \cdot \Omega / b_m = C\Omega / A$$

and with (3.10) and (6.28) the damping of the nutation is

$$\delta_N = -a_{m-1} / b_m + g_{m-2} / g_{m-1}$$

(6.30)
$$\delta_N = -(d_{22} + k_{22}) / A + (k_{22} + Ak_{11} \ m) / C.$$

The damping will remain negative $(\delta_N < 0)$ and the nutation will be stable at high rotor speed if

(6.31) $$A(1 + Ak_{11} / mk_{22}) < C(d_{22} + k_{22}) / k_{22}.$$

This relation is similar to the necessary condition (6.26). It is especially notable that (6.31) shows that the internal damping d_{11} of the transverse motion y_i does not contribute to the

damping of the nutation at high rotor speed. The pervasiveness
of the damping is lost at high speeds.

6.3 Unsymmetrical stiffness and inertia.

The technical effects arising from unsymmetrical stiff
ness and inertia properties and the theoretical difficulties in
calculating them are of such various kind that it cannot be the
aim of a short chapter to deal with them in detail.

These effects are partly well-known and extensive lit
erature exists on it. Some effects and results are reported here.

If an unbalanced rotor with a symmetrical shaft is supported in bearings which have different elasticity coefficients in
transverse directions, then the forced motion of the rotor con-
sists of the superposition of a forward whirl $\left(a\, e^{i\Omega t}\right)$, and a
reverse whirl $\left(b\, e^{-i\Omega t}\right)$. The center of mass, for example does
not whirl on a circle but on an ellypse ($y = a\, e^{i\Omega t} + b\, e^{-i\Omega t}$, y
being its radial displacement) that can degenerate to a straight
line or whirl in any direction. [10, 18]. If the elasticity coefficients are equal the forward whirl only occurs. This may ex-
plain why in experiments, forced reverse whirls can often be
observed even though they are not very distinct, while theoretical investigations assuming symmetric properties of the system
have to say no to their existence.

The unequal stiffness of the shaft in transverse direction lead to equations with coefficients which depend harmonical

ly on time. The stability of motion can thus be seriously affect ed. In simple cases the equations can be transformed into those with constant coefficients when the displacements and inclinations are investigated in a coordinate system rotating with the rotor velocity [1, 3, 11, 18, 27].

An unequal inertia with respect to lateral axes, for example for a two-bladed helicopter rotor, leads as well to e- quations with harmonically varying coefficients [23, 25, 29].

The stability can be studied by means of perturbation theory in order to obtain analytical results for the main insta- bility region or it can be calculated numerically by extending Floquet's theory and deriving from it stability charts [11, 28].

References

[1]   Smith, D. M. : "The motion of a rotor carried by a flexible shaft in flexible bearings"; London, Proceed. Royal Soc., A 142 (1933), p. 92 - 118;

[2]   Biezeno, C. B. and R. Grammel : "Technische Dynamik"; 2. Aufl. Springer-Verlag, Berlin/Göttingen/Heidelberg, 1953;

[3]   Den Hartog, J. P. : "Mechanical Vibrations"; 4. Aufl. McGraw-Hill Book Co., New York, 1956,

[4]   Magnus, K. : "Gyro-Dynamics"; CISM-Publications, Udine, 1970;

[5]   Magnus, K. : "Theorie des Kreisels und der Kreiselgeräte"; Springer-Verlag, Berlin/Göttingen/Heidelberg, erscheint 1971;

[6]   Müller, P. C. : "Special problems of gyrodynamics"; CISM-Publications, Udine, 1970;

[7]   Merkin, D. R. : "Gyroscopic Systems" (in russian), 1964;

[8]   Forbat, N. : "Analytische Mechanik der Schwingungen"; VEB-Verlag Technik, Berlin, 1966;

[9]   Klotter, K. : "Technische Schwingungslehre", 2. Aufl., Bd. 2. Springer-Verlag, Berlin/Göttingen/Heidelberg, 1960;

[10]   Traupel, W. : "Thermische Turbomaschinen", 2. Bd. Springer-Verlag, Berlin/Göttingen/Heidelberg, 1958;

[11]   Schweitzer, G. , W. Schiehlen, P.C. Müller, W.
       Hübner, J. Lückel, G. Sandweg und R. Lauten-
       schlager : "Kreiselverhalten eines elastisch
       gelagerten Rotors"; Ing. Archiv, erscheint
       1971;

[12]   Ziegler, H. : "Kritische Drehzahlen unter Tor-
       sion und Druck"; Ing. Archiv 20 (1952), S.
       377-390;

[13]   Schweitzer, G. und P.C. Müller : "Theoretical
       and experimental optimization of a high-speed
       rotor"; presented at the 1970 ASME Winter
       annual meeting, New York, Nr. 70-WA/Aut-
       11;

[14]   Drenick, R.F. : "Die Optimierung linearer Regel-
       systeme"; Verlag R. Oldenbourg, München
       und Wien, 1967;

[15]   Müller, P.C. : "Die Berechnung von Ljapunov-
       Funktionen und von quadratischen Regelflä-
       chen für lineare, stetige zeitinvariante Mehr-
       grössensysteme"; Regelungstechnik 17 (1969),
       S. 341-345;

[16]   Tondl, A. : "Some problems of rotor dynamics";
       Chapmann G. Hall, London 1968;

[17]   Harris, C.M. and C.E. Crede, ed. : "Shock and
       Vibrations Handbook"; Vol. 3, McGraw-Hill
       Book Company, New York, 1961;

[18]   Dimentberg, F.M. : "Flexural vibrations of rota-
       ting shafts"; Butterworths, London, 1961;

[19]   Kellenberger, W. : "Stabilität rotierender Wellen
       infolge innerer und äusserer Dampfung"; Ing.
       Archiv 32 (1963), S. 323-340;

[20]   Bolotin, V. V.: "Nonconservative Problems of the
         Theory of Elastic Stability"; The Macmillan
         Comp. New York, 1963;

[21]   Mingori, D. L.: "A Stability Theorem for Mechan
         ical Systems with Constraint Damping"; J.
         App. Mech., June 1970, S. 253-258;

[22]   Mingori, D. L.: "Stability of Whirling Shafts with
         Internal and External Damping", to appear in
         Int. J. of Nonlin. Mech., 1971;

[23]   Plainevaux, J. E.: "Mouvement d'un rotor asymé-
         trique tournant dans les paliers élastiques d'
         une équilibreuse", ZAMM 40, (1960), Heft 7/8
         S. 359-367;

[24]   Yamamato, T.: "On critical speeds of a shaft sup-
         ported by a ball bearing", Journal of Appl.
         Mech., Trans. ASME, June 1959, p. 199-204;

[25]   Brosens, P. J. and Crandall, S. H.: "Whirling of
         Unsymmetrical Rotors", J. Appl. Mech., Sept.
         1961, S. 355-362;

[26]   Crandall, S. H. and Brosens, P. J.: "On the Stabili
         ty of Rotation of a Rotor with Rotationally Un
         symmetric Inertia and Stiffness Properties",
         J. Appl. Mech., December 1961, S. 567-570.

[27]   Hemming, G., Schmidt B. and Wedlich Th.: "Erzwun
         gene Scwingungen beim Resonanzdurchgang",
         VDI-Bericht, Nr. 113, 1967;

[28]   Schiehlen, W. und Kolbe O.: "Ein Verfahren zur
         Untersuchung von linearen Regelsystemen mit
         periodischen Parametern", Regelungstechnik
         15 (1967), S. 451-45;

[29]   Weidenhammer, F.: "Parametererregte Schwin-
         gungen ausgewuchteter Rotoren", ZAMM
         1966, T. 145.

Printed in the United States
By Bookmasters